'Invaluable reflections and next-step ideas ... ably the most inspiring and important m... century.'

– **Mike Berners-Lee, author of** ***There is No Planet B: A Handbook for the Make or Break Years***

'Everyone should read this book. It is short, and impassioned, but full of important information: above all an honest and practical plea for us to seize the moment for change urgently. It is inspirational. Please, for all our sakes, read it, and take its timely words to heart.'

– **Iain McGilchrist, author of** ***The Master and his Emissary: The Divided Brain and the Making of the Western World***

'Rupert Read and Samuel Alexander take us deep inside the debates, tactics, and passion that have bound XR together from its founding days, bringing us radical reflections from the frontlines of rebellion. If you want to understand the movement that is finally waking us up, read this book.'

– **Kate Raworth, author of** ***Doughnut Economics: Seven Ways to Think Like a 21st-Century Economist***

'The ecological emergency is the greatest challenge that humanity has ever faced, and Extinction Rebellion may be our last best chance to address it. Read this inside story of the most important social movement of our time.'

– **David Loy, author of** ***Ecodharma: Buddhist Teachings for the Ecological Crisis***

'From the eruption of XR in our lives in late 2018, Rupert Read has been closely involved in the organisation as an advisor, influencer, spokesperson, and occasional critic. These fascinating essays read like dispatches from the front line, crackling with urgency, tempered by timely reflections, and reminding us of the scale of the challenge ahead as we rebuild our shattered, post-coronavirus economies.'

– Jonathon Porritt, former Director of the UK Sustainable Development Commission and of Friends of the Earth

'Activist-philosopher Rupert Read, a key thinker of Extinction Rebellion, has collected a treasure trove of foundational essays, documenting the transformational XR UK experience, that will be immensely valuable for activists around the world trying to replicate that achievement. *Extinction Rebellion: Insights From the Inside* is a keen look at what worked, and what didn't, in the UK, combining on-the-spot observations and public documents from the two years since XR was launched with fresh analysis of how the global movement for ecological sanity may respond in this new COVID-19 world. The book is incisive, pertinent, self-critical, well-written, and, in the XR way, occasionally cheeky. Anyone concerned with combating the broader wave of ecological collapse – that led to the pandemic and that is rising behind it – should read Read.'

– Ken Ward, protagonist in the documentary *The Reluctant Radical*

'Most of us like to watch movies about people saving the world but too rarely do we even try to do it ourselves. Despite the situation so obviously calling for it! This is a book about what might well turn out to be the most important social movement in history.'

– David Graeber, author of *The Democracy Project: A History, a Crisis, a Movement*

'In a world-changing movement that has been careful not to allow itself to get captured by any ego-driven individual "leaders", there is a vital role to be played by those exercising wise thought-leadership in the service of the collective. One of those is Rupert Read. He believes that this civilisation is finished and that the only way that we can end it without violence and without collapse is through the kind of transformation that XR calls for and pre-figures. This eco-philosophy has been vital for "XR 1.0". This book is a great document of that journey, and of how XR 2.0 will need to be if we are going to make it together to a better future. Can XR – can we – deliver on that promise? The human future may hang on the answer.'

– Carne Ross, author of *The Leaderless Revolution* and former diplomat

EXTINCTION REBELLION

INSIGHTS FROM THE INSIDE

Essays by Rupert Read
Edited and with a postscript by Samuel Alexander

EXTINCTION REBELLION
INSIGHTS FROM THE INSIDE

Published by the Simplicity Institute, 2020.

www.simplicityinstitute.org

Copyright 2020 Rupert Read and Samuel Alexander

Cover image based on work by Chaz Maviyane-Davies

Cover design by Andrew Doodson

Layout and typesetting by Sharon France (Looking Glass Press)

Typeset in Bebas and Gill Sans

All rights reserved.

ISBN: 978-0-6488405-1-0 (paperback)

ISBN: 978-0-6488405-2-7 (e-book)

Dedicated, with love and rage, to all the rebels, now and of the future.

CONTENTS

PREFACE

ON EXTINCTION REBELLION 1.0

Extinction Rebellion (XR, for short) is an emergency response. Politics as usual, governments as usual, have let down the peoples of the world in an extreme way: we are on course for eco-driven societal collapse, and we are extinguishing other species very rapidly. We could even end up making ourselves extinct. XR suggests that when your government is driving you and your family over a cliff, it is no longer a legitimate government. Rebellion against it is permitted – indeed, it is *required*. But XR is insistent that such rebellion must be nonviolent. Not only because hurting people isn't nice, but also because there is good reason to believe from the historical record that nonviolence *is frequently more effective* than violence in transforming society. XR asks people to be willing to be arrested, in unprecedented numbers, to leverage such deep, rapid change. Those of us willing to do this are called 'arrestables'.[1] The arrestables are then supported (whether in very practical ways – legally, financially – or simply with vital moral support etc.) by a much larger cohort of non-arrestable sympathisers.

We humans are vulnerable as never before to existential threats resulting from our way of life and in particular from how elites have structured that way of life. In this terrible context, XR has three demands. First, to *tell the truth*. The full truth needs to be told about

1 See Jeremy Harding's very useful article 'The Arrestables' in *London Review of Books*, 16 April 2020, https://www.lrb.co.uk/the-paper/v42/n08/jeremy-harding/the-arrestables

the emergency that we have allowed to be created. Secondly, once that truth is told (by the media, by government, by us all) and understood, it will be possible, as it is necessary, to *act now* to remedy the situation. In concrete terms, this means rapid precautionary, mitigative, and restorative action in the next five years. Rich countries need to lead the way on this: for us to be globally safe, we in such countries need to aim to achieve zero-carbon and zero-biodiversity-destruction by 2025. That is, we need to 'be realistic', as the phrase goes, and 'demand the impossible'. This can be made possible by our unprecedented change-making. Thirdly, how to achieve this eye-watering goal should be decided, not by a representative democracy (which has patently failed), but by citizens themselves: we need *Citizens' Assemblies*. This emergency takes us into a place beyond party politics and beyond ideologies. It is now about survival, and to survive we are going to have to learn a new way of flourishing together. To construct and enact a new, regenerative vision. Assemblies of citizens, well-informed by the best experts, deliberating together, are the best bet for how we could agree to enact the second demand, of acting now, fast and deep enough to save the future.

Why did I choose to work with Samuel Alexander to bring together this material on XR? And why now? There are several reasons:

1. The story of XR is a great story. I am not one of the co-founders who dreamed up XR, but I've been involved since before XR emerged onto the public stage. When I watched Gail Bradbrook's epochal video 'Heading for Extinction and What to Do About It', it reflected back to me many of the things I had been saying in talks in the previous couple of years about how desperate our situation had become. What was different was that this newly forming group, XR, had a plan for what to *do* about it. This gave me a glimmer of real hope for the first time in years. I tracked Gail down, phoned her that same day, and, after a wonderful, lengthy conversation, I threw myself into the embryonic movement. I have tried to serve the movement in various ways, from those beginning days onward. I co-organised and co-wrote the

multi-authored letters that were XR's first ventures into the public domain, which are included at the start of this book. I co-MC'd the launch of XR with Gail Bradbrook in Parliament Square on 31 October 2018, including from a soapbox in the middle of the street outside parliament – our first major direct action, when hundreds of us blocked access to parliament.

This book offers an account of my personal journey, as someone who has found their life's vocation in public-facing, strategic, and political roles within XR since before its launch, and whose life has been completely changed (and completely enriched, made more meaningful and more hopeful) by it. This book doesn't pretend to be anything more than my story inside XR, and therefore it tells very little of the inspiring story of XR outside the UK, although the insights offered will be relevant to XR more broadly). The XR story is a great and necessary one, and I hope that my 'take' on it as laid out here may prove of some interest.

2. I hope it may also prove of *use*. There is a huge amount to be learnt, I believe, from the success of XR, which catapulted real, rapid change in public attitudes to climate and ecology in the UK and to some extent around the world, as well as in the areas corresponding to XR's three demands. If XR itself and other movements can benefit by understanding XR's story somewhat better, then my own telling of that story may be of real use.

3. Now is a great moment to look back at the first two years of XR UK – the first full phase of the movement: *XR 1.0*. Covid-19 marks a historically decisive inflection point in the history of emergency activism, and indeed in the history of humanity. Our previously under-acknowledged common vulnerability to existential threats is now present, lived, acknowledged. The vulnerability 'story', focally present in much of this book and especially in the appendix, 'Rushing the Emergency, Rushing the Rebellion?' – the story that needs to land with the majority of us if we are to have any chance at all of not crashing

civilisation – now finally has a real following wind. Like the virus itself, it has suddenly leapt from the periphery to centre stage. The final chapters in this book document the moment of this virus's epic, deadly, and scary arrival in our lives – and dare to look beyond its dominance. If XR can resonate with the felt vulnerability that has traversed the world in the last weeks and months, that will matter more than any direct 'actions' it can take at the moment.

I write from my home in Norwich, England, in late spring 2020. Under coronavirus lockdown. But I can communicate with Sam in Australia, and with others all over the globe. The shared experience of vulnerability, of existential threat, that is tearing through many humans and communities, especially in the 'Global North', for the first time in the living memory of most of us, is something which, if it can be extended to other longer, tougher emergencies, will change everything. Thus, in the midst of the awful, avoidable tragedy of Covid-19, I have more hope than I have had for years about the awful, avoidable mega-tragedy of climate and ecological breakdown. The birth of XR gave me hope when I had had virtually none; paradoxically, the corona crisis is refuelling that hope. XR from now on will be very different, for a host of reasons implicit in what I have just written: right now, because of the domination of public attention by the virus and because locking down makes most direct action impossible; and, as we go forward, because, after this enforced pause everything can be different, as I describe in my final chapters in this book. At the time of writing, XR UK had recently decided temporarily to pause most of its physical activities, during the height of the corona crisis; wisely, in my view. All of this adds up to why many of us are speaking of this as the moment of birth (or, at least, of conception) of *XR 2.0*. Which will be a different story, suitable for telling years from now. Assuming, as I hope and believe but cannot take for granted, that most of us will be (t)here to tell that story.

4. Last but not least, XR needs money: now more than ever, in a more stretched, poorer world whose attention has been dragged (and not wrongly!) to something nearer at hand. After covering costs, all proceeds I make from this book will go to XR UK (and part of XR UK's income goes to XR International). If I can help leverage a little cash for this amazing movement, which has achieved so much on a shoestring compared to established NGOs, then so much the better.

I want to offer now my deep thanks to those who have co-authored certain of the pieces brought together in this book with me: Jem Bendell, John Foster, Marc Lopatin, Skeena Rathor, Dario Kenner, Frank Scavelli, Alison Green, and Richard House. Thanks too to the far larger group who have inputted into many of these pieces in one way or another; and to those hosting the venues where some of them were first published. Thanks to Atus Mariqueo-Russell, Treve Nicol and Josie Wilson for assistance in helping research some of these pieces. Deep thanks finally to Antoinette Wilson, a superb copyeditor, and of course once again to Sam Alexander, a brilliant interlocutor and guiding editorial presence.

This book is written in service to the cause of averting extinction(s), and in humble awareness of the far larger organism of which I am but one interwoven strand. This book obviously had no possibility of existing at all without the willingness of thousands upon thousands of rebels to be arrested in this greatest, most necessary, hardest of causes. If I can see anything clearly as attested in these pages, it's because of these Extinction Rebels and the many thousands more standing behind them. If you are one, then thank you. (If you are not, then perhaps you might just be, after reading this book.) I stand with you all, arms locked together virtually for now and in reality again one day soon. I stand with you and I salute you.

– Rupert Read, Norwich, England, May 2020

INTRODUCTION

BATSHIT-CRAZY TIMES

Samuel Alexander

If there is one thing a virus cannot kill, it is a rebellion. Nevertheless, there is no way to begin this collection of essays on Extinction Rebellion, at this time, other than by acknowledging the remarkable, mind-bending moment at which these words are written. It is a time of pandemic, one destined to shape the future of human civilisation for years if not decades ahead. In Australia, from where I write, the economy has all but shut down, with little open for business besides medical centres and hospitals, supermarkets and food outlets, and a very select number of other essential services. Against every ideological bone in its body, our conservative government has announced unprecedented stimulus packages, to avoid masses of people in our affluent nation from falling into destitution.

Because so many people have lost their livelihoods from this unexpected suspension of capitalism, banks have had to freeze mortgage repayments for six months and rental evictions are currently prohibited. The national and state borders have been closed, and public gatherings of more than two people are prohibited. Someone in Australia was recently fined for eating a kebab on a park bench. All of this was unthinkable a few months ago. Today it is normal.

This is the stuff typically reserved for dystopian fiction, not real life, but many other nations around the world are in a similar position to

Australia, with more destined to follow as the Covid-19 virus continues its extraordinary disruption. I look at the date on my computer and see that it is April 1st, usually a time for jokes and pranks. I hesitate for a moment: is this for real? Surely someone is playing us for fools. But this is no joke. These are truly batshit-crazy times and I certainly won't pretend to fully understand what is happening, and none of us could know what lies ahead. For all I know I am writing from within a relative calm that could yet prove to be the eye of an even more transformative hurricane. For now, all I can do is nod approvingly at the words degrowth scholar Jason Hickel recently cast out into the Twitterverse: 'Capitalist realism is over. Everything is thinkable.'

Indeed it is. At the same time, tragic though the pandemic is, we need to remember that Covid-19 is a crisis within a broader ecological and humanitarian crisis. It was only three months ago when my home nation was ablaze, suffering a devastating fire season drawing international attention, owing to conditions that were exacerbated by global heating. It is estimated that over one billion animals perished in the furnaces – one billion! Who has the emotional capacity to understand that statistic? And what does next summer bring for our shared Anthropocene? What the coronavirus shows, however, is that we really can act as if the house is on fire, when we feel it is urgent enough.

Therein, of course, lies the catch: *when we feel it is urgent enough.* Does it follow that our politicians and the dominant culture did not yet realise that climate change, species extinction, topsoil erosion, deforestation, pollution, resource depletion, population growth, poverty, and inequality were real problems of equal or great urgency? Regrettably, that seems to be the case. The world knew enough about the science of climate change in 1988 to establish the Intergovernmental Panel on Climate Change (IPCC). And yet, last year carbon emissions continued to rise, over thirty years after the IPCC was established to warn and guide us.

In the midst of the current pandemic, which is causing so much human suffering, it is clear that shutting down the aviation industry and much of consumer culture is allowing a moment for the planet to take pause from the onslaught of global industrialism. For so long we have been told that it was not possible. And yet here we are, albeit by disaster not design. As French philosopher Bruno Latour recently commented: 'Next time, when ecologists are ridiculed because "the economy cannot be slowed down," they should remember that it can grind to a halt in a matter of weeks worldwide when it is urgent enough.' In particular, let us remember this capacity to swiftly downshift the economy, should critics assess Extinction Rebellion's demand of decarbonising by 2025 and casually dismiss it as non-viable. It will certainly be non-viable if we do not try.

Still, if national and international public discourse is anything to go by, it seems the primary goal of politics in this time of disruption is to 'bounce back' to where we were before the pandemic. Of course, all the evidence suggests that bouncing back would be no solution at all. We must not bounce back. We must bounce otherwise and elsewhere. The key question, then, is: bounce back to where and how?

What is most disturbing about the current pandemic is how quickly everything else gets erased from our attention. In Australia, the media has all but forgotten about the bushfires. How short our memories are. But these issues – climate destabilisation and all the rest – have not disappeared. We live in the same Earth system, only one currently under the duress of a virus. As if we did not have problems enough already! For the moment, perhaps, people have even forgotten about Extinction Rebellion. But if there is one thing a virus cannot kill, it is a rebellion.

* * *

At base, Extinction Rebellion is about mobilising people for collective action, with the goal of producing a just and sustainable world through various practices of nonviolent civil disobedience. Given that Covid-19 has meant that Extinction Rebellion is, for the time being, unable to take to the streets in rebellion, Rupert Read and I decided to continue our participation by working from our homes on this book.

It is my honour to be editing this book of Rupert's essays on Extinction Rebellion and contributing this introduction and an extended post-script essay, 'The Rebellion Hypothesis'. Rupert is an activist-academic with the Department of Philosophy at the University of East Anglia, UK. That may position him as an unlikely figure to feature so prominently in one of the most disruptive environmental movements on the planet right now. After all, philosophers aren't normally the ones taking to the streets in rebellion. But then again, philosophers in the past have encouraged us not merely to interpret the world, but to change it. It is a lesson that Rupert has taken to heart.

Even to call Extinction Rebellion a 'movement' is somewhat misleading. It is a rebellion – a global rebellion, fast becoming a movement of movements. Currently it is simmering in cyberspace during the physical distancing enforced by the pandemic, waiting to remerge, I suspect, with more energy than ever. It will be interesting to see how the Citizens' Assemblies upheld by Extinction Rebellion will address the key issues of this moment in history – this turning point, perhaps – including: how to manage protective contraction of the economy in the future, and how best to build a more resilient, just, and sustainable society in the wake of Covid-19. In this book you will learn much about the movement's history and mission, and some thoughtful reflections on where it may move next.

As indicated in the preface, we offer this collection of Rupert's writings as an 'insider's perspective' on the first two years of this fast-emerging and evolving rebellion. One person's angle on this movement will inevitably be limited by the narrowness of personal experience. But that narrowness can also be enriching and full of insight. No doubt by the time this book goes to print the story of Extinction Rebellion will have undergone further twists and turns, with more in store. That is how it will be, and that is how it should be. So watch this space, or better yet, help shape it.

While I don't always agree with everything Rupert says, for years I have been part of a growing audience enriched by his provocative writings, lectures, interviews, and public talks. His work on Extinction Rebellion is particularly good, justly receiving a huge amount of attention, with qualities that transcend the content, namely, the qualities of honesty, clarity, and depth.

You too may come away from this book with questions or criticisms of Rupert's work, but he would both invite and celebrate this critical engagement. Indeed, it is one of the primary reasons we publish this book: to provoke further debate and discussion of Extinction Rebellion and related movements. Our goal is certainly not to get everyone to think the same thing. Far from it. The goal is to foster and exploit the diversity of human ingenuity by fuelling the fires of public discourse, in the hope that this better enables us all to respond to the variety of civilisational problems that will demand many knowledges, insights, and practices if they are to be tolerably addressed.

Beyond his writing and scholarship, Rupert deserves credit for walking the talk in a way that is challenging in the best sense of the term. It seems that at every opportunity Rupert has put down his metaphorical pen and left his desk to participate in the various uprisings of Extinction Rebellion (primarily those near his base in Norwich). I am sure this active engagement is part of the reason why his writing and talks have had the reach and impact they have: they are authentic.

My formal collaboration with Rupert began with the publication of our book *This Civilisation is Finished: Conversations on the End of Empire and What Lies Beyond*, published in English by the Simplicity Institute in 2019 (and currently under translation into German, Spanish, and French). That book was a collection of conversations between us that explored a wide array of issues, ranging from future scenarios for globalised capitalism, pessimism over technological solutions, alternative conceptions of progress, political activism (including Extinction Rebellion), the role of a teacher in a dying civilisation, among many other 'big picture' topics.

To our surprise the book quickly moved thousands of copies, which is unusual for books by academics. This cultural reception suggests to us that there is a growing desire for societal commentary that aspires to truth-telling without censorship or sugar-coating. You will find the same spirit of uncompromised honesty shaping the following pages.

This book chronicles the period from before XR's launch to the final phase of 'XR 1.0', marked by the coming of the coronavirus; that is, starting in summer 2018 and going to Spring 2020 inclusive (Northern hemisphere). And towards the end of the book Rupert starts to look forward to XR 2.0, reflecting on how the experience of shared vulnerability and emergency from the coronavirus might reshape the historic task of XR: waking the world up at last to the true gravity of the climate and ecological emergency, and shifting from warm words to sufficient action.

The chapters of this book speak well enough for themselves, so there is no need for an extended overview of the content, and each chapter is introduced by Rupert with a few sentences of context. While the book was designed to be read from front to back, readers should feel free to consult the Table of Contents and dip into the book and jump around as they see fit. For now, suffice to say that this anthology collects together the wide range of thinking and writing Rupert has shared over the last couple of years, as he has sought to understand and guide Extinction Rebellion. Many have been published before, some are appearing here for the first time. The diverse contributions include introductory essays, longer reflections, short articles or pamphlets, transcripts of interviews, conversations with other activists and scholars, theses on Covid-19, among other topics and forms. We bring them together here for convenience and provocation. But, of course, this is not the end of the story. It is one perspective on, and experience of, the story's beginning. The future of Extinction Rebellion is unwritten. This book is an invitation to help write it.

* * *

As the Covid-19 pandemic deepens or exacerbates the range of pre-existing crises, it seems that our collective task now is to ensure that these destabilised conditions are used to advance progressive humanitarian and ecological ends, rather than exploited to further entrench the austerity politics of neoliberalism. I recognise, of course, that the latter remains a real possibility, as did the arch-capitalist Milton Friedman, who expressed the point in these terms:

> Only a crisis – actual or perceived – produces real change. When that crisis occurs, the actions that are taken depend on the ideas that are lying around. That, I believe, is our basic function: to develop alternatives to existing policies, to keep them alive and available until the politically impossible becomes the politically inevitable.

I'm not often in complete agreement with Milton Friedman, an ideological nemesis, but on this point I am. For those who recognise the potential in this moment to think and act differently, our basic function is to keep hopes of a radically different and more humane form of society alive, until what today seems impossible or implausible becomes, if not inevitable, then at least possible and perhaps even probable. It is through crisis that citizenries can be sufficiently perturbed that the sedative and depoliticising effects of affluenza and apathy might be overcome.

Indeed, I cautiously suggest that it is better that citizens are *not* in fact protected from every crisis situation, given that the encounter with crisis can play an essential consciousness-raising role, if it triggers a desire for and motivation towards learning about the structural underpinnings of the crisis situation itself. I believe social movements, including Extinction Rebellion, should be preparing themselves to play that educational and activist role, and in fact it is heartening to see this already unfolding in the many inspiring social responses to this time of pandemic.

While Covid-19 is currently demanding physical distancing, Extinction Rebellion continues to organise online. More broadly, during these turbulent times we see communities being there for each other, in mutual aid and support. And, in the wake of Covid-19, Extinction Rebellion will be back on the streets en masse, because our work is not yet done. Until then, and beyond, we hope you find this book fuel for the fire of ecological democracy.

CHAPTER 1

I'M AN ACADEMIC EMBRACING DIRECT ACTION TO STOP CLIMATE BREAKDOWN

This piece was published in autumn 2018, in The Conversation.[2] *It serves I think as a useful introduction to why so many of us decided to take the life-changing step of getting seriously involved in XR. I sought to write in a relatively 'neutral' way; readers might nonetheless sense strongly, almost as I was writing the piece, the pull of the cause.*

Extinction Rebellion is a nonviolent direct action movement challenging widespread inaction over dangerous climate change and the mass extinction of species which, ultimately, threatens our own species. XR demands a just and democratic transition to a better, safer future.

Saturday 17 November 2018 is 'Rebellion Day', when people opposed to what they see as a government of 'climate criminals' aim to gather together enough protesters to close down parts of the capital – by shutting down fossil-powered road traffic at key pinch-points in London.

Our long-term aim is to create a situation where the government can no longer ignore the determination of an increasingly large number of people to shift the world from what appears to be a direct course

2 https://theconversation.com/extinction-rebellion-im-an-academic-embracing-direct-action-to-stop-climate-change-107037

towards climate calamity. Who knows, the government could even end up having to negotiate with the rebels.

As someone who is both a veteran of nonviolent direct actions over the years and an academic seeking to make sense of these campaigns, I've been thinking quite a lot about what's old and what's new about Extinction Rebellion. Here are my conclusions so far.

FROM WORLD PEACE TO CLIMATE JUSTICE

Extinction Rebellion is rooted in longstanding traditions exemplified by the radical nuclear disarmament movement. The founders of Extinction Rebellion have thought carefully about past precedents, and about what works and what doesn't.

They've noted, for instance, that you don't necessarily need active involvement from more than a tiny percentage of the population to win radical change, provided that you have a noble cause that can elicit tacit backing from a much larger percentage.

Extinction Rebellion is also quite different from its predecessors. True, the disarmament movement was about our very existence, but nuclear devastation was – and still is – only a risk. Extinction Rebellion's aim is to prevent a devastation of our world that *will* come – and quite soon – unless we manage to do something unprecedented that will radically change our direction.

Climate activists often compare their struggle to victories from the past. But in my view comparisons which are often made – to Indian independence, the civil rights movement, or the campaign for universal suffrage, for example – are over-optimistic. These historical movements were most often about oppressed classes of people rising up and empowering themselves, gaining access to what the privileged already had.

Extinction Rebellion challenges oligarchy and neoliberal capitalism for their rank excess, and the political class for its deep lack of commitment, beyond warm words, to serious action on the ecological emergency. But the changes that will be needed to arrest the collapse of our climate and biodiversity are now so huge that this movement is concerned with changing our whole way of life. Changing our diet significantly. Changing our transport systems drastically. Changing the way our economies work to radically relocalise them. The list goes on.

This runs up against powerful vested interests – but also places considerable demands upon ordinary citizens, especially in 'developed' countries such as the UK. It is therefore a much harder ask. This means that the chances of Extinction Rebellion succeeding are relatively slim. But this doesn't prove it's a mistaken enterprise – on the contrary, it looks like our last chance, our best hope.

This all leads into why I sat in the road blocking the entrance to Parliament Square on 31 October, when Extinction Rebellion was launched – and why I will be 'manning the barricades' again on 17 November. I cherish the opening words of the famous Shaker hymn 'Simple Gifts': "Tis the gift to be simple'. What does it mean to live simply at this moment in history? It means to do everything necessary so that others – most importantly our children (and their children) – can simply live.

It isn't enough to live a life of voluntary simplicity.[3] One needs also to take peaceful direct action to seek to stop the mega-machine of growth-obsessed corporate capitalism that is destroying our common future. That's why it seems plain to me that we need peaceful rebellion now, so that we and countless other species don't face devastation or indeed extinction.

The next line of that Shaker hymn goes, "Tis the gift to be free.' In our times, to be free means to not be bound by laws that are consigning our children to purgatory or worse. If one cares properly for one's

3 See my article on this co-authored with Sam Alexander and Jacob Garrett: 'Voluntary Simplicity: Backed By All Three Main Normative-Ethical Traditions', https://ueaeprints.uea.ac.uk/id/eprint/67014/1/Accepted_manuscript.pdf .

children, that must entail caring for their children, too. You don't really care for your children if you damn their children. And that logic multiplies into the future indefinitely – we aren't caring adequately for any generation if the generation to follow it is doomed.

As mammals whose primary calling is to care for our kids, it is therefore logical that an outright existential threat to their future, and to that of their children, must be resisted and rebelled against, no matter what the pitifully inadequate laws of our land say.

I've felt called upon to engage in conscientious civil disobedience before, at Faslane and Aldermaston against nuclear weapons and with EarthFirst in defence of the redwood forests threatened with destruction in the Pacific Northwest of the USA.

But Extinction Rebellion seems to me the most compelling cause of them all. Unless we manage to do the near impossible, then after a period of a few decades at most there won't be any other causes to engage with. It really is as stark and as dark as that.

If you too feel the call, then I think you now know what to do.

CHAPTER 2

FACTS ABOUT OUR ECOLOGICAL CRISIS ARE INCONTROVERTIBLE: WE MUST TAKE ACTION

Alongside Alison Green and Richard House, I co-wrote and co-organised this letter that appeared in The Guardian *in October 2018,*[4] *signed by ninety-four academics, including Professor Joy Carter, Vice-chancellor, University of Winchester; Dr Rowan Williams, former Archbishop of Canterbury; Prof. Danny Dorling; Dr Mark Maslin; Prof. Jason Hickel; Prof. Jem Bendell; Dr Ian Gibson, Former Chair, House of Commons Science and Technology Select Committee; Dr Susie Orbach; Dr David Drew MP, Shadow minister, environment, food and rural affairs; and Professor Molly Scott Cato MEP.*[5] *This was the first entry of XR onto the national scene, setting the stage for our launch event and first national direct action, five days later. It was significant that these academics (including a good number from climate and ecological sciences) were willing to stand up in public in favour of civil disobedience for this cause.*

4 https://www.theguardian.com/environment/2018/oct/26/facts-about-our-ecological-crisis-are-incontrovertible-we-must-take-action?CMP=Share_iOSApp_Other

5 The full list of signatories can be found here: https://www.theguardian.com/environment/2018/oct/26/facts-about-our-ecological-crisis-are-incontrovertible-we-must-take-action?CMP=Share_iOSApp_Other. The accompanying in-depth article 'We Have a Duty to Act' made, in fact, an even bigger splash, as this was the first journalistic investigation into the plans of Extinction Rebellion, immediately prior to the launch of XR at Parliament Square: https://www.theguardian.com/environment/2018/oct/26/we-have-a-duty-to-act-hundreds-ready-to-go-to-jail-over-climate-crisis. That article features first-person mini profiles of Gail Bradbrook, Alison Green, Jessica Townsend, and myself.

We the undersigned represent diverse academic disciplines, and the views expressed here are those of the signatories and not their organisations. While our academic perspectives and expertise may differ, we are united on one point: we will not tolerate the failure of this or any other government to take robust and emergency action in respect of the worsening ecological crisis. The science is clear, the facts are incontrovertible, and it is unconscionable to us that our children and grandchildren should have to bear the terrifying brunt of an unprecedented disaster of our own making.

We are in the midst of the sixth mass extinction, with about two hundred species becoming extinct each day. Humans cannot continue to violate the fundamental laws of nature or of science with impunity. If we continue on our current path, the future for our species is bleak.

Our government is complicit in ignoring the precautionary principle, and in failing to acknowledge that infinite economic growth on a planet with finite resources is non-viable. Instead, the government irresponsibly promotes rampant consumerism and free-market fundamentalism, and allows greenhouse gas emissions to rise. Earth Overshoot Day (the date when humans have used up more resources from nature than the planet can renew in the entire year) falls earlier each year (1 August in 2018).

When a government wilfully abrogates its responsibility to protect its citizens from harm and to secure the future for generations to come, it has failed in its most essential duty of stewardship. The 'social contract' has been broken, and it is therefore not only our right, but our moral duty to bypass the government's inaction and flagrant dereliction of duty, and to rebel to defend life itself.

We therefore declare our support for Extinction Rebellion, launching on 31 October 2018. We fully stand behind the demands for the government to tell the hard truth to its citizens. We call for a Citizens' Assembly to work with scientists on the basis of the extant evidence and in accordance with the precautionary principle, to urgently develop a credible plan for rapid, total decarbonisation of the economy.

CHAPTER 3

ACT NOW TO PREVENT AN ENVIRONMENTAL CATASTROPHE

Six weeks after the Guardian *article, Alison Green, Richard House, and I co-ordinated another multiple-signatory letter in* The Guardian, *this time with a global focus.*[6] *This letter expressed support for XR and broadly allied movements across the world, especially those movements already non-violently fighting hard for nature and a future in the Global South. Signatories included Vandana Shiva, Naomi Klein, Noam Chomsky, Philip Pullman, Bill McKibben, Tiokasin Ghosthorse, Jonathon Porritt, Lily Cole, Salim Dara, Chris Packham, David Graeber, Giorgos Kallis, and Kate Raworth.*[7]

In our complex, interdependent global ecosystem, life is dying, with species extinction accelerating. The climate crisis is worsening much faster than previously predicted. Every single day two hundred species are becoming extinct. This desperate situation can't continue.

Political leaders worldwide are failing to address the environmental crisis. If global corporate capitalism continues to drive the international economy, global catastrophe is inevitable. Complacency and inaction in Britain, the US, Australia, Brazil, across Africa and Asia all illustrate

6 https://www.theguardian.com/environment/2018/dec/09/act-now-to-prevent-an-environmental-catastrophe?CMP=Share_iOSApp_Other

7 For the full list of signatories, see https://www.theguardian.com/environment/2018/dec/09/act-now-to-prevent-an-environmental-catastrophe?CMP=Share_iOSApp_Other

diverse manifestations of political paralysis, abdicating humankind's grave responsibility for planetary stewardship.

International political organisations and national governments must foreground the climate-emergency issue immediately, urgently drawing up comprehensive policies to address it. Conventionally privileged nations must voluntarily fund comprehensive environment-protection policies in impoverished nations, to compensate the latter for foregoing unsustainable economic growth, and paying recompense for the planet-plundering imperialism of materially privileged nations.

With extreme weather already hitting food production, we demand that governments act now to avoid any further risk of hunger, with emergency investment in agro-ecological extreme-weather-resistant food production. We also call for an urgent summit on saving the Arctic icecap, to slow weather disruption of our harvests.

We further call on concerned global citizens to rise up and organise against current complacency in their particular contexts, including indigenous people's rights advocacy, decolonisation, and reparatory justice – so joining the global movement that's now rebelling against extinction (e.g., Extinction Rebellion in the UK).

We must collectively do whatever's necessary non-violently, to persuade politicians and business leaders to relinquish their complacency and denial. Their 'business as usual' is no longer an option. Global citizens will no longer put up with this failure of our planetary duty.

Every one of us, especially in the materially privileged world, must commit to accepting the need to live more lightly, consume far less, and to not only uphold human rights but also our stewardship responsibilities to the planet.

CHAPTER 4

AFTER THE IPCC 1.5 DEGREES REPORT: WAKING UP TO CLIMATE REALITY

I published this piece on Medium[8] *in the run-up to XR's launch. It is the main piece in which I set out my points of divergence from Jem Bendell, author of the widely read 'Deep Adaptation' paper. While Jem's work has been formative for some in XR, I am nervous that its effect, contrary to his intent, has sometimes been to blunt clarity about the need to follow the precautionary principle – to prevent, to 'mitigate' in the technical sense of that term as it is used in climate-policy. The nuanced stance I seek to uphold may be not only a little more accurate (because it is less 'knowing' about what we don't actually know, in a world sometimes full of surprises and beyond our ken) but also more psychologically efficacious.*

The exciting – but also terrifying – 'Global Warming of 1.5°C' special report from the IPCC made a few headlines; and now the reporting has mostly moved on. The mega-story of potential #climatebreakdown, the long emergency that threatens to *overwhelm* us, the news story that should be on our screens every night, has been overtaken by dramas in Brussels and Westminster (not to mention on Strictly Come Dancing).

8 https://medium.com/@rupertread_80924/after-the-ipcc-report-climatereality-5b3e2ae43697

The IPCC report is highly *conservative* about the dangers we face[9] and dares not challenge capitalism or 'growth'.[10] Consequently, it sketches only the most basic elements of the unprecedented transformation the world must undertake. But an increasing number of voices are being raised to argue that it is no longer enough to stake everything on that transformation being achieved. Indeed, it would be a rash person who would *bet* on it *being* achieved, for the IPCC report *is*, if anything, considerably too optimistic. A soberer view has been presented in Green House Think Tank's 'Facing Up to Climate Reality' project,[11] and by my friend and colleague Professor Jem Bendell. His recent paper 'Deep Adaptation' has gone viral, making an impact that is surely a sign of the times.

Jem's paper, which makes the striking and scary claim that our civilisation will collapse due to climate-degradation within ten years of the time of its writing has plainly struck a chord. It is waking people up, allowing them to share the fears that they have been having privately for, in many cases, years now.

Taking as given the welcome impact that Jem's paper is having, I'll take a moment to spell out here the two key differences between my own thinking and his. These differences may sound small, but I think they are significant enough to be worth dwelling on. In part, because they may make a significant difference to how the Extinction Rebellion message is received.

1. Jem claims that societal collapse is now 'inevitable'. I think collapse is not certain, but 'only' *almost* inevitable. In my view it makes no sense to make hard predictions about a system of which one is a part, for what happens depends in part on what we do. And we

9 Compare, by contrast, Schellnhuber's frank view about the dangers here in the strikingly titled 'I Would Like People to Panic': https://horizon-magazine.eu/article/i-would-people-panic-top-scientist-unveils-equation-showing-world-climate-emergency.html & also Wolfgang Knorr's here: https://jembendell.com/2019/07/31/climate-scientist-speaks-about-letting-down-humanity-and-what-to-do-about-it/

10 See Prof. Kevin Anderson's take on this point, here: http://blog.policy.manchester.ac.uk/posts/2018/10/response-to-the-ipcc-1-5c-special-report/

11 https://www.greenhousethinktank.org/facing-up-to-climate-reality.html

don't know (the limits of) what we are capable of until we try. In principle it is still possible for us to escape the dire future that awaits us if we attempt merely reform rather than a profound compassionate revolution.[12] The point about it being in principle impossible to predict the human future, because of our agency, is very important, for it keeps open a space of freedom and faith and courage that a 'doomer' message risks closing down; the 'It's inevitable' message risks turning us into observers rather than actors. I know that that is the opposite of the intention of Bendell's exciting 'Deep Adaptation' agenda, but it *is* nonetheless a risk that some will find that message of inevitability disempowering, undermining the open potentiality of humanity.

2. Jem claims that collapse is 'near-term', and gives an upper bound for it of a decade from now. I think we can't know that collapse is near-term, because of the reason already sketched in (1), but also because we have to be humble and accept the limits of our knowledge. To think that we can know when collapse will come is to make the same kind of mistake of hubristic over-confidence in predictions that is commonly made in the mainstream among scientistic thinkers over-confident in their models. My work in recent years on precaution has convinced me of this. It is hubristic to claim that we can know the future. It is, as I say, exactly a symptom of how our society has got itself into such trouble. The point, as I've stressed in my joint work on this with Nassim Taleb,[13] is that we have to get used to living as safely as possible in a world that we do not 'fully' understand (and never will). This has radical implications for how we change the way we are living. We should seek to decreate fragile systems, and reduce our dependency on predictions.

12 Here in 'Apollo-Earth', a piece I co-authored with Deepak Rughani of Biofuelwatch, is a sketch of one way in which such a better future remains perfectly possible: https://theecologist.org/2017/mar/09/apollo-earth-wake-call-our-race-against-time

13 See, e.g., our 'The Precautionary Principle': https://www.fooledbyrandomness.com/pp2.pdf. See also https://www.greenhousethinktank.org/precautionary-principle.html for some less technical takes.

I think that we need to be wary of hostages to fortune. If in 2028 we are somehow still standing, then people will come back and refer negatively to Jem's 'ten years' semi-prediction. (Remember the treatment meted out to the 'Limits to Growth' pioneers.) It will be used to discredit him/us.

(1) and (2) are inter-related points. We should be very careful how we handle them, however. Unless we are strong-willed and determined to keep facing up to climate reality, we might think that they deliver a 'reprieve'; that we can then put aside the strong medicine that Jem is prescribing. That would be a drastic mistake, and the last thing I want. Neither (1) nor (2) undermines the centrality of the Deep Adaptation agenda to what is now needful. They only complicate the picture, make it a little more uncertain in application, take it further from the dangerous certainties of the 'doomer' or the survivalist.

For let me stand shoulder to shoulder with Jem in saying that the future looks extremely grim – unless we somehow manage to transform our entire way of life beyond recognition, rapidly. The situation is particularly grim in the Arctic. The albedo loss there is highly disturbing, threatening in itself to blow the IPCC scenarios away, as Jem details. And above all there is the possibility that there may be a methane time-bomb.[14] If the staggeringly vast amounts of methane buried below now-thinning ice and 'permafrost' start to be 'liberated' then we will be not looking 'only' at the end of human civilisation, but at the possible extinction of humanity and of most animals. Conceivably, we might become 'committed' to that outcome even within a decade.

What I think *we can know* is that this civilisation is finished. We don't quite know for certain that it will end in collapse, and we don't know how long until it is finished, but we can be fully confident that it will not survive in anything even remotely resembling its current form. For that form is cancerous. If our civilisation survives then it will have utterly transformed. It will no longer in any meaningful sense be this civilisation. (I elaborate on my belief that this civilisation is finished in chapters 8 & 9.)

14 See, e.g., Peter Wadhams's take here: 'Arctic Research and the Methane Risk', https://www.youtube.com/watch?v=D3L0R6LzEUE

Again, I think that outcome – transformational adaptation, an utter, rapid changeover to an ecological civilisation – is deeply, obviously desirable. And again, I'm very sceptical that it will be achieved. If that is my thinking, how, therefore, should I act? If you find my line of thinking convincing, how should *you* act?

That's the 64 trillion dollar question, to which this book as a whole offers a kind of answer. Clearly, a central part of that answer (and one to which Jem assents) is: to rebel. Another part of it is us at every level (household, community, society, globe) starting to act more precautiously. The beauty of the precautionary principle approach is that we don't need to make predictions about 'inevitability' or about a specific time period. The logic of precaution points us in the correct direction anyway, whether the chances of collapse are 3%, 33%, or 100%; whether it is likely in three months, in three years, or in thirty years. (My own rough best guess right now is that we are facing a very severe but probably non-total collapse that will unfold over a generation or so.[15])

I think that this precautionary logic may be more helpful to our cause than the standard scientific 'evidence-based' logic that Jem attempts to extrapolate from – a logic that is so pervasive in the rhetoric of our world now, obsessed as it is by the image of science,[16] but a logic that is actually often harmful. *I believe that my way of characterising our situation is more likely to be energising and motivating* than Jem's. Extinction Rebellion is probably a last chance for us to do enough to stop climate catastrophe, or at the very least to very significantly slow it. But it risks being undermined by a message that says near-term social collapse is inevitable.

Once again, to be clear, while I'm a tiny bit more optimistic than Jem, the emphasis needs to be on the word tiny. The crucial thing is to

15 For more detail, see my article 'Some Thoughts on Civilisational Succession': http://www.truthandpower.com/rupert-read-some-thoughts-on-civilisational-succession/

16 For full buttressing of this claim, see Jonathan Beale and Ian James Kidd (eds): *Wittgenstein and Scientism*, Routledge, London, 2019), including my essay therein, on how to avoid scientism in relation to climate.

start taking seriously the strong likelihood, unless together we make something miraculous happen, that collapse is coming, and probably *not* in the distant future. So while still seeking to enable the miraculous, and feeling liberated by the direness of the emergency from norms of politeness, law-abidingness, etc., we must also have a plan B. We have to start talking about and preparing for the probability of failure. We have to start, for example, trying to make nuclear waste relatively safe against a future in which our governmental institutions will not be there to prevent it melting down or catching fire. It would be plain reckless, at this point, to bet everything on our pre-empting collapse.

And that is still a very fearful message. So my final point is: let's create a place where that fear can be felt, voiced, shared. One of the most powerful things we can do right now is share not our predictions or our precautions but our emotions.

I'm afraid. For my students. For all our children. For my loved ones. For myself.

When one bonds emotionally about this with those one is communicating with (especially if they are younger than one) then that is powerful. That is the power of the so-called 'powerless'. That's the truth I want to speak.

I'm *scared*, dear reader. Not 'just' for future generations. For you, and for me. Join me, in this honest fear.

It is time to share our fears – and to rebel, to seek to stop them being realised.

CHAPTER 5

VENICE: A CANARY IN THE CLIMATE COALMINE

When Venice flooded just days before the launch of Extinction Rebellion, I was invited to contribute this short piece to The Independent.[17]

31 October 2018 – All Hallows' Eve – is the day of the launch of 'Extinction Rebellion'. Hundreds of us will be descending upon Parliament Square, to declare that we're no longer willing to stand by while the powers that be frogmarch our species (and most wildlife) towards extinction. Climate chaos has become an existential threat. And so we're rebelling.

As if to show that God has no sense of humour, He's unleashed record-matching floods in Venice to accompany the occasion. Three quarters of the city is underwater. This is tragic. And just a tiny taste of the vast tragedies to come, if we don't act convincingly to match the global emergency we find ourselves in.

Actually, this has precious little to do with God, and everything to do with humanity. Increasingly, 'global weirding' means that 'natural disasters' aren't natural at all: they aren't 'acts of God'. They're caused by us. One of the main three factors behind Venice's increasing susceptibility to floods

17 https://www.independent.co.uk/voices/venice-flooding-italy-weather-climate-change-environment-civil-disobedience-a8609116.html

is climate-change-caused sea level rise. (Another is offshore methane extraction – itself a big contributor to dangerous climate change.)

Venice is a uniquely beautiful place; a wonder of the modern world. I remember with great pleasure attending a conference there. It's bizarre to see pictures of places one has strolled through now a metre under water.

Eventually Venice will succumb. It will become unliveable, and then gradually sink beneath the waves. An early casualty of climate devastation. Like Bangladesh. Like the Maldives. Unless, just conceivably, we act in ways that are completely outside the box.

In the 1990s, I joined EarthFirst! for 'Redwood Summer' in California, and we put ourselves on the line to try to stop the clearcutting of those magnificent trees. In the 2000s, I was part of the movement seeking through civil disobedience to disrupt the Trident nuclear missile system; and I interrupted proceedings in the House of Commons, to protest against the use of cluster bombs in Iraq. (I ended up spending an afternoon in the cells at the Palace of Westminster: yes, they have their own tiny prison there.) But it's been years since I engaged in non-violent direct action.

I don't particularly want to be arrested. I don't want my life disrupted. I'm no wannabe hero.

It's simply that we cannot wait any longer, while the planet burns. Our own government, in its infinite unwisdom, used the Budget yesterday – just a fortnight after the IPCC's compelling report on our planetary emergency – to throw £30,000,000,000 at road-building while one five-hundredth of that amount went to tree planting, to 'compensate'. Truly shameful, at this moment in history.

So I'm willing to risk arrest.

The idea of Extinction Rebellion is that, rather than allowing ourselves to be gradually extinguished, we will insist upon the government changing course radically, now. And if they refuse, some of us are ready to go to prison for it.

We may well fail. The odds are stacked against us. The latest blow being the election of a genocidal,[18] ecocidal[19] extremist to the presidency of Brazil.

And we can take little comfort from previous successes. The civil rights movement, or women's suffrage, or gay marriage: these were liberations for visible groups of oppressed human beings, and didn't challenge the basic economic system or the exploitation of nature. What we are trying to do instead is to save the future for all of us; and that means changing everything.

Any rational person would bet heavily against us. But to be able to look our kids – or indeed Venetians – in the eye, we need at least to try. Imagine how you'd feel in a decade's time, once it's too late – if you hadn't even tried.

And even if we fail, perhaps we'll have slowed the juggernaut of destruction down some. Even that would be better than nothing. Much, much better.

There's still time. Why not join us? It's the hour of decision, the eve of destruction, before we become ghosts, our civilisation at best a memory. And who knows, given the time of year, maybe some of us will even be wearing fun masks…

We've little to lose by throwing caution to the winds in our own lives. For we're already losing Venice; and that's just the canary in the climate coal mine. When globally we stand to lose everything, we may as well get serious about stopping it…

So, let's try.

18 See, e.g., https://theintercept.com/2019/02/16/brazil-bolsonaro-indigenous-land/ for evidence to support this serious claim.

19 See, e.g., https://www.theguardian.com/environment/2019/aug/25/g7-cant-turn-blind-eye-to-amazon-ecocide-forest-fires-indigenous-tribes

CHAPTER 6

EXTINCTION REBELLION BEYOND LONDON

This article, published in The Ecologist *in December 2018,*[20] *suggests that it is dangerous for XR to focus its energy on actions that seem to target ordinary working people. This is a theme I returned to in 2019 in my pamphlet 'Truth and its Consequences' (Chapter 13 of this book), and in early 2020 in a pamphlet co-authored with Skeena Rathor and Marc Lopatin, 'Rushing the Emergency, Rushing the Rebellion?' (the Appendix to this book). It should be noted, as indeed I do in 'Truth and its Consequences', that the April 2019 Rebellion proved me wrong to some extent: we achieved success then by forcing a national conversation that changed the game on climate and ecology, by (among other things) inconveniencing ordinary people. But I submit that that was a one-off, a superbly surprising success that caught the authorities and in fact everyone (including ourselves!) off guard, something that by its very nature could only happen once.*

The ante-upping direct-action intimated in the final paragraph of this piece took place and was a great success, partly because it targeted the local 'democratic' elite. We shut down Norfolk County Council for several hours, and temporarily stopped them from being able to fund the building of the road discussed below.[21]

20 https://theecologist.org/2018/dec/10/extinction-rebellion-beyond-london?fbclid=IwAR1t-JScpAqYQRXDfQRkv-zuhzflLr0erSKxUayipP0ymj5BlWBWezWvuztQ

21 See these links for details: https://www.youtube.com/watch?v=GCqK5aFDDkc&feature=youtu.be https://www.bbc.com/news/uk-england-norfolk-46439210 https://www.youtube.com/watch?v=3tK_FniBVnQ&feature=youtu.be . At time of writing Norfolk County Council is still trying to go ahead with building the road, and it remains unclear whether they will actually be able to go ahead and do so.

Extinction Rebellion has rapidly made a name for itself – by way of unleashing an unprecedented scale of nonviolent direct action (NVDA) in London.

We launched on 31 October 2018, by blocking vehicular access to parliament. On 17 November, 5000 of us blocked five bridges across the Thames. The first phase of direct-action protests came to a head with 'Rebellion Day 2' on 24 November 2018, in which we marched on Downing Street and Buckingham Palace.

Meanwhile, the movement is already internationalising. But what next for XR in the UK?

XR is starting to facilitate actions everywhere. Of course, the thing about the climate is that it is under threat by all manner of human activities. Most obviously, the way we grow our food, what industry does, and the way we travel.

COMMON FUTURE

I have thrown myself headfirst into this movement nationally – and also locally in my home city of Norwich.

Our long-term aim is to create a situation in which the government can no longer ignore the determination of an increasingly large number of people to shift the world from a direct course towards climate calamity. That will only happen if the movement causes trouble everywhere, not just in London.

So, in places like Norwich – and I predict you will soon see the same happening across the country – some of us have started putting our bodies on the line for the sake of our common future.

SHAM CONSULTATION

Last week, Norwich XR undertook its first NVDA. The councils in Norfolk are determined to build a truly appalling new road, across a river that is a Site of Special Scientific Interest (SSSI) and a Special Area of Conservation. This road would of course be a contribution to increasing our nation's carbon emissions at the very time we need to slash them. And it would threaten to further extinguish the area's biodiversity.

Norfolk County Council is running a sham consultation, trying to get the public to fixate on which route should be built across the Wensum, rather than on whether the road should be built at all. It is quite obvious that building a new road like this is the height of absurdity, even insanity, at a time when the UN – which is actually highly conservative in such matters – is telling us that we need to halve our carbon emissions within a decade.

We occupied the consultation exhibition space in the centre of Norwich. We presented passers-by with genuine information instead, explaining why this road is the worst of all the dreadful road projects that have been proposed and built in Norfolk, and in particular why its climate-irresponsibility is absolute and unacceptable.

We effectively shut down the 'consultation' for three-and-a-half hours. There were about fifty of us. Too many to easily deal with or intimidate. So, although the managers of the space threatened to call the police on us to force us to leave, in the end they appeared not to have done so. We were able to blockade the 'consultation' for the whole time.

LEGITIMATE TACTICS

I was pleasantly surprised by the positive reaction we garnered from the vast majority of passers-by. I think one reason is that we were not inconveniencing them (apart from a tiny handful who actually wanted to see the council's rubbish materials about the road 'options' – but we let them through). We were inconveniencing the powers that be.

I'd like to draw a general moral from that. I believe that XR actions ought to target politicians (local and national), civil servants, the authorities, big business, the very rich – not ordinary people.

Sure, shutting down roads is a perfectly legitimate tactic, because transport emissions are killing us, and still rising. But it is not usually a very effective tactic if it annoys ordinary people. If we close down government departments, local councils and parliament, executive offices, carbon-polluting factories, and the like, we are far more likely to keep the broad mass of the public with us, while we civilly disobey.

Those who took part in this NVDA in Norwich, especially the many newbies to this game, were emboldened.

NEXT ACTIONS

We already have a plan in mind to up the ante against this awesomely terrible road building plan whose absurdity we've put on the map.

The groundswell of XR is already beyond expectations – and it has certainly moved well beyond London. If we are strategically and tactically smart, and keep many people on side, we will radically subvert the powers that be and the fossil economy. We might even win…

CHAPTER 7

ADULTS MUST NOT OBJECT TO THEIR CHILDREN TAKING RADICAL ACTION

I was asked to contribute this piece to The Conversation *in February 2019,*[22] *on the occasion of a multi-authored letter by UK academics (many of them – though probably by no means all – backers of XR) in support of the school climate strikers.*

A worldwide wave of school climate strikes, begun by the remarkable Greta Thunberg, has reached the UK. Some critics claim these activist-pupils are simply playing truant, but I disagree. Speaking as both a climate campaigner and an academic philosopher, I believe school walkouts are morally and politically justifiable.

Philosophy can help us tackle the question of whether direct action is warranted via the theory of civil disobedience.[23] This states that, in a democratic society, one is justified in disobeying the law only when other alternatives have been exhausted, and the injustice being protested against is grave.

In the case of the climate school strikes, it is without question that the injustice – the threat – is grave. There is none graver facing us.

22 https://theconversation.com/school-climate-strikes-why-adults-no-longer-have-the-right-to-object-to-their-children-taking-radical-action-111851

23 For an explanation of the theory, see: https://plato.stanford.edu/entries/civil-disobedience/ See also, this book's post-script essay by Sam Alexander.

It appears reasonable to claim furthermore that other alternatives have indeed been exhausted. After all, people have been trying to wake governments up to the climate threat for decades now, and we are still as a society way off the pace set out even by a conservative organisation such as the IPCC.

But if that claim were strongly contested and it were suggested that climate activism should continue to focus on conventional electoral politics, then attention might revert to the assumed premise that society is democratic. Do people in Britain and elsewhere really live in 'democracies', given (for instance) the vastly greater power of the rich and of owners of media to influence elections, compared to everyone else?

I don't want to adjudicate whether we really live in a democracy. But what of course makes this a particularly salient question for school strikes is the simple fact that in any case children have no voice in this democratic system. And yet the climate crisis and the perhaps equally catastrophic biodiversity crisis will affect children much more than adults.

Our 'democratic' system seems to have a built-in present-centric bias and a concomitant weakness in relation to issues of long-term significance, that seriously undermine its claims to democratic legitimacy. Thus philosophers have sometimes argued, beginning with Edmund Burke in the eighteenth century, that to make the system truly democratic we would need to somehow include – and give real power to – the voices of the past and the future in that system. Most especially, for they are at risk of suffering the worst, the voices of children and indeed of unborn future generations.

So, a forceful argument could be made that it must be legitimate for children to take part in climate actions, for they do not even have recourse to the democratic channels (such as they are) that adults take for granted. This is especially true once we add that it seems reasonable for children to object to schooling that may well be rendered irrelevant by a climate-induced catastrophe. For example, much of the way that economics, business studies, and IT are taught presupposes a world that will probably soon cease to exist.

ADULTS HAVE FAILED

If you are convinced by this, then all well and good. However, at this point, I want to pull the rug slightly from under the argument that I've made so far. I put it to you that, if you are an adult, as I am, then your view in any case is somewhat beside the point.

For the brutal fact is that, try hard though some of us have done, we adults have categorically failed our children. This is a grievous wrong, perhaps the worst thing that mammals, primates, such as ourselves, can do: to have let down those whom we claim to love more than life itself. We have set our children on a path to a 'future' in which society as we know it may have collapsed. And even if we accomplish an unprecedented societal transformation over the next decade, the massive time-lags built into the climate system mean things will still get worse for a long time to come.

And so on this occasion we adults ought to humbly realise that it is no longer for us to tell our children what to do. We ought rather to take up the role of supporting them in their uprising, asking how we can help them in their struggle for survival. They are inspiring us, now.

The ultimate reason why we should support these school strikes, as I and hundreds of other UK academics have just declared we will do, is because we adults – through the inaction which has created the present desperate global situation – have forfeited the moral right to do anything else.

CHAPTER 8

WHY I SAY THIS CIVILISATION IS FINISHED

This previously unpublished piece connects the book you're reading with my previous book, co-authored with Sam Alexander, This Civilisation Is Finished *(2018). That book, and my lecture of the same name given at Cambridge University on which the initial concept of it was based (a talk which went viral on YouTube in the early months of XR), have sometimes been misunderstood as 'giving up'. As I set out here, nothing could be further from the truth.*

It has been a huge privilege to be involved with Extinction Rebellion. For the first time in years I feel a growing glimmer of hope for humanity. Finally, we are seeing a mass mobilisation of people who are not willing to die quietly. An upwelling of people unafraid to call for the radical initiatives that we need to limit the scope of global overheating. As a spokesperson for Extinction Rebellion, I have been among those privileged to put the case for the action of our rebels to those in the media and in government, as well as to the public.

We need to be clear that there is no 'safe' level of warming of the planet. Even preventing 2 degrees (which is now almost unachievable) means the death of over 99% of the world's coral reefs – permanently defacing the ecology of our planet – and probably means the end of ice in the northern hemisphere. The International Panel on Climate Change – which is still, contra popular belief, a relatively conservative body – is unambiguous in its latest report, 'Global Warming of 1.5°C', that 2 degrees means the displacement of millions of people through desertification and flooding. It means a much greater frequency and a higher magnitude of the extreme weather events that are increasingly blighting the world. It means an increase in violence and war globally because of resource scarcity and hotter temperatures. It is violence: 2 degrees *is* violence from the rich and stupid against the global masses. It means increased frequency of pandemic and pestilence, with greater threats to our health and the food supply we rely upon to nourish us. And because of the inherent unpredictability of the effects of 2 degrees warming, it could expose us to a myriad of other threats that we cannot predict and that could be far worse than current models suggest.

This is why Extinction Rebellion's actions are so important, and in particular why the call for net zero UK emissions by 2025 is vital. Our movement has been courageous by communicating with brutal honesty exactly what is at stake over the climate emergency. There needs to be far more of this communication within the public sphere.

In my recent book, *This Civilisation is Finished,*[24] co-authored with Samuel Alexander, we attempt exactly this. We reject the 'soft denialism' so often present in the mainstream discourse about the climate emergency. A discourse that seems cherry-picked to present what is actually ecological apocalypse in as palatable and unthreatening a way as possible. Instead, we have found that minds and hearts are only truly concentrated when the scale and enormity of the threat to human and non-human life is exposed in its unveiled magnitude. When this occurs, people stare the threat in the face, the fight-or-flight response is activated and – as there is nowhere to run – they become energised by the necessity to battle for the survival of themselves and their children.

This is no exaggeration. The stakes of course are very, very high, because the climate crisis and the broader ecological emergency puts the whole of what we know as civilisation at risk. By 'this civilisation' I

24 *This Civilisation is Finished*, Rupert Read and Samuel Alexander, Simplicity Institute, Melbourne, 2019.

mean the hegemonic civilisation of globalised industrial growth capitalism – sometimes called 'Empire' – which today governs the vast majority of human life on Earth.

As I see things, there are three broad possible futures that lie ahead:

1. *This civilisation could collapse utterly and terminally* as a result of climatic instability (leading for instance to catastrophic food shortages as a probable mechanism of collapse), or possibly sooner than that, through nuclear war, pandemic,[25] or financial collapse leading to mass civil breakdown. Any of these are likely to be precipitated in part by ecological/climate instability, as Darfur and Syria were.

Or

2. *This civilisation (we) will manage to seed a future successor-civilisation(s)* as this one collapses.

Or

3. *This civilisation will somehow manage to transform itself* deliberately, radically, and rapidly, in an unprecedented manner, in time to avert collapse.

The third option, the aim of XR, is by far the least likely, though the most desirable, simply because either of the other options will involve vast suffering and death on an unprecedented scale. In the case of (1), we are talking the extinction or near-extinction of humanity. In the case of (2) we are probably talking at minimum multiple megadeaths. But (2) would obviously be hugely preferable to (1), and thus the ultimate importance for us of getting our societies not only to mitigate but also to *adapt*, deeply.

25 This was written before the onset of Covid-19. This new coronavirus is a black swan in its specificity, but a white swan in terms of broad predictability. We always knew there would be a new mega-pandemic, especially given our witless destruction of biodiversity etc. The only question was exactly when, where, and what.

The second option is very difficult to envisage clearly, but is, I now believe, very likely. Unless we are incredibly lucky or incredibly determined and brilliant (or more likely both) then we are facing, almost certainly, changes around the world which are going to bring an end to this civilisation. So we need to think about what comes after it. We need to think about it now, and we need to start to work towards it; because there are many sub-possibilities within possibility two, and some of them are very ugly.

One of the reasons Sam and I wrote our previous book was to talk about how we can prepare the way for (2). I think that there has been criminally little of that preparation to date. Virtually everyone in the broader environmental movement has been fixated on the third option, unwilling to consider anything less. I strongly believe now that that stance is no longer viable. And, encouragingly, I am not quite alone in that belief.

The first option might soon be as likely as the second. It leaves little to talk about.

Any of these three options will involve a transformation of such extreme magnitude that what emerges will no longer in any meaningful sense be *this* civilisation: the change will be the kind of extreme conceptual and existential magnitude that Thomas Kuhn, the philosopher who coined the term 'paradigm shift', calls 'revolutionary'. Thus, one way or another, this civilisation is finished. It may well run in the air, suspended over the edge of a cliff, for a while longer. But it will then either crash to complete chaos and catastrophe (Option 1); or seed something radically different from itself from within its dying body (Option 2); or somehow get back to safety on the cliff-edge (Option 3). Managing to do the latter miraculous thing would involve such extraordinary and utterly unprecedented change, that what came back to safety would still *no longer in any meaningful sense be* this civilisation.

That, in short, is what I mean by saying that this civilisation is finished.

Extinction Rebellion is key to transforming the civilisation we have into something that will allow us to maintain human life either in the third option or in arming our global consciousness with the understanding of the need for Deep Adaptation in the face of the second option.

If not, we are left only with terminal collapse.

I hope that this book, a kind of sequel to the previous book that Sam and I put together, and a book in which obviously I am discussing XR at some length, will help us in these difficult but necessary thought-and-feeling processes. I hope that you, dear reader, will be willing to use all nonviolent means necessary to seek to avoid Option 1. Let's end this failed civilisation wisely and voluntarily, not in chaos and terminal collapse.

CHAPTER 9

DANGEROUS ANTHROPOGENIC CLIMATE CHANGE AND DEEP ADAPTATION

This piece was published in the Ecologist *in February 2019.*[26] *It draws upon and summarises an academic working paper, 'So what is to be done?',*[27] *published as IFLAS's Occasional Paper number 3 (Jem Bendell's epochal 'Deep Adaptation' paper had been Occasional Paper number 2 in that series). The version published here has been edited to avoid repetition of matters already discussed.*

I want to start out by addressing younger readers in particular, and what I have to say to you is stark. It is this: your leaders have failed you; your governments have failed you; your parents and their generation have failed you; your teachers have failed you; and I have failed you.

We have all failed to raise the alarm adequately, and so of course we have failed to prevent the dangerous climate change that is now here, and the worse climate change that is coming and that is definitely going to get a lot worse still – definitely, because of time-lags built into the system.

This crisis already shows our failure. For if we had been going to tackle this in such a way as to actually get a grip on it, we would have done so a generation ago – at minimum.

26 https://theecologist.org/2019/feb/08/climate-change-and-deep-adaptation

27 http://lifeworth.com/IFLAS_OP_3_rr_whatistobedone.pdf

TRUE LEADERSHIP

Roughly speaking, we would have elected Green or genuinely green-friendly, non-growth-obsessed governments everywhere in the world a generation ago and they would have done things that were quite unpalatable to a lot of us. That would have been true leadership.

But of course, nothing remotely like this happened. So now we're in a real last chance saloon. The globally hegemonic civilisation of which we are all a part is in an endgame. Those who wanted to preserve it have already definitively failed.

Because of that failure, I'm afraid for you, reader, especially if you are younger than me (I'm 52). I fear that some of you are unlikely to grow old.

We've gambled too much on succeeding in preventing or at least mitigating anthropogenic dangerous climate change and the anthropogenic extinction crisis because we were unwilling to face up to the alternative. But the alternative is not as simple as an instantaneous end of life would be. 'The' alternative is in truth complex, multiform. It involves many possible variants of 'unthinkably' horrendous, bad – and even (in some respects) good.

TRANSFORMATIONAL ADAPTATION

Most crucially, there is a huge difference between the various versions of complete irrecoverable societal/species collapse, on the one hand, and the rise of a successor civilisation(s) out of the wreckage of this one, on the other.

We have to be willing to think this – and face it. Which means that we have to look beyond mitigation alone; we have to get serious about the processes of *transformational adaptation* and *deep adaptation* that are now necessary.

We cannot continue to avoid the vast effort necessary in attempting to adapt our communities to cope with our changed and changing world. Not least because the time-lags built into the climate system mean that, even in the extraordinarily unlikely event that we manage to stop massively damaging our climate further, it is bound to deteriorate further for a long time to come.

The only way that our civilisation might appear to persist is if we manage to transform it beyond recognition. But that transformed civilisation would then in no meaningful sense be the same civilisation as ours.

It would be *radically* relocalised, degrowthist, energy-descended; it would have ended consumerism and foregrounded ecology; it would have re-learnt indigenous and peasant wisdom and left behind most of the wrong turn of industrial capitalism; in short, it would probably be as different from our present world as we are from the world before the industrial revolution.

IRREVOCABLE CHANGES

The great task that now lies ahead of us is the work of taking seriously the effort of adaptation to our irrevocably changed world. Deep Adaptation, as discussed in the previous chapter, means adaptation premised upon collapse. And it has to be faced plainly that such collapse is likely.

For instance, how many more summers like the one of 2018 can we take? In my own neck of the woods, in Norfolk, many crop yields were massively down. And this is while we deal with the effects of only 1 degree of global over-heat. What will things be like when we reach 2 degrees, or even 3, which is now a realistic expectation?

EXTINCTION?: REBELLION...

I have argued (see Chapter 4) that Jem Bendell's claim that we face 'inevitable', 'near-term' social collapse is not valid. But we must bend our wills to Deep Adaptation, as an insurance policy against the *likelihood*, nevertheless, of collapse. And to transformational adaptation, adaptation that seeks simultaneously to mitigate and to transform our society in the direction it desperately needs to change. The latter points, too, towards the hope for that transformation, a hope that remains, even in the darkness of this time.

Insofar as human beings are willing to wake up and to look the dark reality of climate crisis in the eye, so we rise up to meet it. That is true courage. That is still at the heart of the task now upon us. A task that Extinction Rebellion is leading the way towards.

CHAPTER 10

HOW CLIMATE GRIEF MAY YET BE THE MAKING OF US

I was asked to contribute this piece to The Conversation[28] *in August 2019 at the sombre moment of the first recorded 'death' of a glacier in Iceland, which the people of that country, famously aware of their own history and identity, wisely decided to mark. This article makes the case that grief comes from love, and that grief can motivate; that climate grief (and eco-grief more generally), therefore may be our best chance of coming to do* enough. *As we in XR are desperately seeking to do.*

Congratulations to the Icelandic people, a people who have held on fiercely to a sense of their heritage and cultural identity during times when other cultures have been too ready to let these go. What is the latest example of this pride in heritage? Iceland has commemorated the loss of the Okjökull glacier with a poignant plaque. This is radically new. And it is something we will encounter more often, tragically, in the hot new world we have created: an awareness and commemoration of the parts of nature that our climate-recklessness has eliminated. We might see this plaque as a potent example of an emerging phenomenon: climate grief.

28 https://theconversation.com/iceland-is-mourning-a-dead-glacier-how-grieving-over-ecological-destruction-can-help-us-face-the-climate-crisis-122071

Before I delve into the nature of our grief at our decimation of nature, a decimation that threatens now to end our civilisation, let me rehearse the philosophy of 'ordinary' grief, as typified by grief at the death of a loved one.[29]

Grief is how love survives loss. Grief is a reaction of pain and even outrage against the ripping of someone (or something) precious from out of one's lived world, the tearing of the very fabric of that world.

Our very world has to change for us to accommodate the death of a loved one, and for us in due course to emerge from grief. Denial in this context, correctly understood, is not necessarily an irrational or delusive belief; it is rather the initial form of (highly painful) transition from one lived world to another. If there is in due course to be complete acceptance, then, strange as it sounds to say it, there must first be some denial. For only some such denial points adequately towards the extreme depth of the loss in question. The loss being so extreme, denial is tacitly a way of acknowledging its full depth: the lost one was part of the ground of one's very world, rather than a substitutable element in it.

Denial need not then be the opposite of acceptance; it is a potent transitional *means* to it. (One might say, instead: the true opposite of acceptance is professed indifference.)

Grief radically differs in its logic from (ordinary) sadness over a loss. If an acquaintance dies, or if a loved object is lost, this does not imply the need for one to construct a new or renewed 'world' in response. Deep grief by contrast amounts to a deformation of one's lifeworld. Metaphorically, one might describe it as having a hole punched into (or ripped out of) one's lived world. That is why, unlike some small sadness, grief necessitates a transformed lifeworld: to somehow repair or live with that hole.

---

29 For more detail on such grief, see my https://medium.com/@GreenRupertRead/what-is-grief-a-personal-and-philosophical-answer-d83d7f288c96 .

As it is experienced, then, when it really *is* experienced, grief is not the removal of one object among others from the world; rather, the entire *character* or *form* of the world is altered. To put this in a 'gestalt' metaphor: it is a change not in figure but in ground. Sadness is a figure on a world with a secure taken-for-granted ground. Grief involves rather the reconfiguring *of the ground itself* (which takes time).

Grief springs from the depth of our interconnectedness, which could even be called our internal relatedness with one another, or our collective wholeness. Grieving arises because, contrary to the ideology of our liberal individualist society, we are *not* detached from one another.

In grieving, what one does is acknowledge the rip, the tear in that world that the loved one's passing made.

This helps us to understand the less venal forms of climate-denial. It helps us understand, that is, why some feeling, decent, and intelligent people are for a long time tempted into its post-truth absurdities. For denial, properly understood, is a part of grieving; in that it is far too crude to think that losing a person who was very close to one is simply a loss of one substitutable element in the lifeworld. Such loss rather threatens and *qualitatively alters* one's lifeworld. Similarly, it is in a certain sense unimaginable, even absurd, to think of us destroying our very climate. That climate is not one substitutable element in the lifeworld. It is its very ground. No wonder people resist, deny.

To get beyond such denial requires you to remake your very world.

Denial in such a case is, then, the not-altogether-unreasonable resistance to such devastating loss as I've described; the motivated rebellion against it. It is not believable that one has lost the person; it is too awful; the world one took oneself to be securely part of would have to be radically different, in order for one to have done so. This denial may, at first, take the form of a temptation towards out-and-out disbelief; then perhaps of ghostly or spectral presences (the lost loved one is present in their absence); then perhaps of a continued pattern sometimes of more and more sporadic thinking and hoping as if the person is still alive. The lifeworld one inhabits in such ways resists the absence of the non-substitutable one. Similarly, it isn't surprising that so many people have been desperately hoping that the science must somehow be wrong, or acting as if we can still hope for the continuation of our same old world; while burning fossil fuels like there's no tomorrow.

It requires strength to gradually turn denial into acceptance and to build a new life.

How, though, do climate grief and extinction grief differ from grief at the loss of a loved one?

When we 'lose' a species or an ecosystem or some previously-taken-for-granted part of our future, it's actually been murdered ('lost' is a euphemism). Thus, woke eco-grief is typically angry as well as distraught. It resembles the grief of someone close to a murder victim.

But there is a difference even from that case. Climate grief and grief over the degradation of our beautiful natural world never lessen, let alone go away. Because these emergencies will define our entire lifetime, and at present (and without doubt for a long time to come, because of the time-lags in the system) are still getting worse.

Thus, while a healthy reaction to grief over a lost loved one is to grieve deeply and then gradually to 'recover', there is no 'recovery' from ecological grief.

The only recovery from it that is possible at all is for us to change the world such that it no longer keeps deteriorating.

This is how eco-grief, grief at the tearing from us of what we love and are not willing to do without, leads into radical eco-action. When we finally allow ourselves to face the full terrible reality of what our species is doing to our home and our kin and our very future, then, to avoid getting stuck in depression or despair (which are understandable responses), there is no alternative but to struggle. And given how far gone we are now, because we allowed denial to rule for so long, that surely means: to rebel.

This is how grief expresses and powers the love that is the one thing that might yet save our future from being – to vary Orwell – a boot stamping on the faces of all beings, humans included, forever, until perhaps there are few or no such beings left.

There is a vast mental health crisis coming. Those suffering from eco-grief – including, probably, those in Iceland who were willing, at Okjökull, to face the pain of the loss of the very thing that gives their culture its name – are in the vanguard of it. This crisis – a pandemic of grief, depression, and worse that will rise in intensity as the world's citizens wake up to the slow death-march that their 'leaders' and the world's rich and powerful more generally are laying on for them – may yet, however, be the making of us. For what powers rebellion is facing the terrible truth of the decaying future we have manufactured; what enables us to face up to that truth, after we work through denial and depression, is our grief; and that, because grief is an expression, above all, of a profound love.

CHAPTER 11

AN OPEN LETTER TO DAVID WALLACE-WELLS: HOW *THE UNINHABITABLE EARTH* MAY BE TOO OPTIMISTIC

David Wallace-Wells's important book The Uninhabitable Earth *attracted some criticism for being too 'negative'. This piece, written in response to that book, was co-authored with Jem Bendell and John Foster (author of* After Sustainability*). We published it in* The Ecologist *shortly before the April Rebellion in 2019,*[30] *and our argument was the opposite: that the book was too 'positive'. That even Wallace-Wells was not looking deeply enough into the dark reality of our predicament, and that only by doing so do we make it possible for us to despair, to grieve, to re-energise, to realise that we have very little to lose – and to rise up.*

It should be noted that when we spoke of geoengineering here, we meant both Solar Radiation Management (SRM) and Carbon Dioxide Removal (CDR) technologies. The reason why so-called Negative Emissions Technologies (NETs) have been rebranded recently as not-geoengineering is precisely because geoengineering has come to have such a bad rep: and that bad rep is in my opinion justified.[31] *So we did not mean in this*

30 https://theecologist.org/2019/apr/04/open-letter-david-wallace-wells

31 This is laid out in detail in my chapter on all this in the Green House book 'Facing Up to Climate Reality' (which was referenced, via John Foster's review, in our *Ecologist* piece). That chapter can be accessed at https://www.greenhousethinktank.org/uploads/4/8/3/2/48324387/fucr_foster_chapter_6_updated.pdf .

piece to imply that Wallace-Wells was backing SRM (which he does not); 'only' that his backing for CDR (aka NETs) could still be quite dangerous.

The more fundamental point we sought to make here concerned the use by Wallace-Wells of the concept of 'engineering' in a much broader sense to encompass solutioneering/mitigation as a sole means of extricating us from the road to climate catastrophe. For that is an area in which he is significantly more optimistic than the three of us. We believe that a mitigation-focused approach is not honest – not truthful – enough, and that movements like XR need to tell the truth more fully about the extent to which adaptation now must be attempted.

This 'Open Letter' did indeed begin a useful dialogue with Wallace-Wells, including in person at a public event at the XR Tent at the Byline Festival, in August 2019, where we found ourselves in even more agreement than we had previously realised.

Dear David,

As a trio of British academics who think and write about dangerous anthropogenic climate change, we have been impressed by your new book, *The Uninhabitable Earth: A Story of the Future.*[32]

This already best-selling book, like the viral article in *New York Magazine* from which it grew, states with passion and eloquence the hard truths of our current global plight. Far from being irresponsibly alarmist, as some have alleged, your straight look at oncoming disaster offers a vital stimulus to realistic understanding and action.

We are so pleased that your book is receiving the mass attention it deserves, and is thereby making the very real risk of an unprecedented climate breakdown and consequent societal collapse comprehensible to the general public.

32 A *Guardian* review of the book appears here: https://www.theguardian.com/books/2019/feb/17/david-wallace-wells-uninhabitable-earth-review

AGAINST GEOENGINEERING

As one of us has stated in a published review,[33] however, we also fear that your book may lead people to believe that the unprecautionary deployment of geoengineering is the answer to our predicament.

We are unconvinced by your claim that because we *engineered* this mess, so we must be able to *engineer* an escape from it. While that may be a neat journalistic turn of phrase, it is a logical nonsense.

Dangerous climate change was not intentionally engineered by humanity. The self-reinforcing feedbacks that are further heating our world show us how the complex living system of Planet Earth is beyond direct human control. So, we have no precedent for humanity intentionally engineering global change.

We understand you may wish to offer your readers some hope. However, your argument offers a continuing licence for the hubris which has led humanity into climate-peril in the first place.

You point out that since 'a decarbonised economy, a perfectly renewable energy system, a reimagined system of agriculture and perhaps even a meatless planet' are in principle possible, we have 'all the tools we need' to stop tragedy in its tracks. And yet that would require us, as you also sardonically note, to rebuild the world's infrastructure entirely in less time than it took New York City to build three new stops on a subway line.

DEEP ADAPTATION

It is dangerous to hang on to such an unrealistic hope while not making adequate preparations for the likelihood that it will prove groundless.

Really facing up to climate reality, by contrast, means giving up all hope of solutions – without giving up on hope itself.

33 https://www.greenhousethinktank.org/the-uninhabitable-earth.html

Instead of fantasies of one-world command-and-control salvation, we believe that *The Uninhabitable Earth* should wake us all up to the need for what one of us has recently and influentially named a 'Deep Adaptation agenda'.

This involves building resilience, both physical and psychological, learning to relinquish long-held beliefs and aspirations (such as that of uninterrupted 'progress'), and the attempted restoration of attitudes and practices which our carbon-fuelled way of life has so dangerously eroded.

Such an approach, while recognising the certainty that the civilisation which has brought us to this pass is finished, accepts also that we cannot know in advance what fine human and societal possibilities may emerge from the crucible of this very recognition.

TRANSFORMATIVE CHANGE

The irony of your starkly-titled book is that it ends up being, from our perspective, too 'optimistic'. This may blind readers to the greatest new need now: for Deep Adaptation – that is, for accepting that some kind of eco-induced societal collapse is now not merely possible, but likely, and preparing honestly for it; for recognising that – while it is absolutely vital to continue to seek to mitigate our society's climate-deadly emissions – the time is past when it was credible to fixate on doing this while ignoring the increasingly urgent need for Deep Adaptation.

What we draw, and should like others to draw, from your urgently necessary book is a difficult but – we believe – genuinely realistic message of hope.

It is not that acknowledging the hard truths which you present so starkly might still enable us to avoid climate disaster. It is now probably too late for that, as in practice you so clearly demonstrate. Rather, it is the hope that through accepting the inevitability of such disaster for our present civilisation, we may yet find our way to genuinely transformative change, capable of avoiding terminal catastrophe for humanity and the biosphere.

The sooner we realise that humanity won't have a Hollywood ending to human-triggered dangerous climate change, the more chance we have to avoid ours becoming a true horror story.

We invite you to think with us about what facing up to climate reality now really means, and in particular to enter into the Deep Adaptation agenda.

CHAPTER 12

TELLING THE TRUTH: THE IMPLICATIONS OF XR'S DEMAND 1

Thanks chiefly to the heroic efforts of over 1000 arrested Rebels, the April 2019 Rebellion in London achieved astounding success in terms of a public consciousness breakthrough. In my role as part of XR's Political Liaison unit, I requested a formal meeting between XR and the UK Government. Michael Gove, the Environment Secretary, agreed to meet with us. Greta Thunberg sealed the deal by getting him to agree to it publicly .

With assistance from XR colleagues, including Jem Bendell, I prepared this document for that meeting. It is previously unpublished and its fairly raw style left mostly intact. Our team included Clare Farrell, Sam Knights, and Farhana Yamin (Chair). The meeting,[34] attracted the interest of much broadcast media, and set the scene for the debate in parliament the next day. To the surprise of many, Gove and the government did not oppose Labour's motion that day for the declaration of a Climate and Environment Emergency. This declaration was non-binding but represented the first national move toward XR's Demand 1: to tell the truth.

34 A full video record of the meeting can be viewed here: https://www.youtube.com/watch?v=EMGqP5rP8v8

TELLING THE TRUTH ABOUT BRITAIN'S RECORD ON CLIMATE: LET'S STOP LYING WITH STATISTICS

In a statement to the House of Commons responding to the April Rebellion, Claire Perry claimed that the UK has been a world leader on climate.[35] Central to Perry's case for this claim was her specific claim, a very widespread one with which the Climate Change Committee is complicit, that Britain is a 'world leader' because our climate-deadly carbon dioxide emissions have allegedly fallen by around 40% since 1990.

Perry uses the word 'territorial' to describe UK emissions in this context. However, that word conceals the massive and rising emissions produced by air travel, shipping, and the manufacturing of products we buy from countries such as China.

When those emissions are included, UK emissions have hardly dropped since 1990.[36] And, because our contribution in this country to these three areas is larger than that of many other countries, we then suddenly look no better than many other countries (all of which, like us, are way short of where they need to be to avoid civilisational collapse in short order).

We want to urge this government to begin telling the truth on emissions. No UK government ever has. If the present government were to admit that, on a full accounting, Britain's emissions have dropped little since 1990, and that our record, on this accounting, is not very different from that of a number of other 'developed' countries, that would be a huge step forward towards Demand 1, of telling the truth. That would be the first step of true leadership in regard to climate.

Messaging around climate does not have to be 'optimistic'. In fact, messaging around climate that tells the truth about our justified feelings of fear and grief as we face this existential crisis has made the world's

35 https://www.theyworkforyou.com/debates/?id=2019-04-23a.600.10

36 https://kevinanderson.info/blog/capricious-foes-big-sister-high-carbon-plutocrats-irreverent-musings-from-katowices-cop24/

greatest ever climate mobilisation, over the last two weeks, an incredible success.

In this foundational case of the emissions stats, then, telling the truth means sea, air, and embodied emissions must all be included in headline UK climate data going forward. Doing so would remove the basis for the dangerous complacency caused by the misleading message that we are 'world leaders' on climate action. It would also undercut the highly dubious meme, which discourages climate action here, that 'China's emissions matter far more than ours do', because in this regard we *are* China. If we hadn't offshored so many emissions to China, then much air and shipping emissions would have been saved, and less-energy-efficient industry in China wouldn't have been encouraged.

TELLING THE TRUTH ABOUT THE DIRE THREAT WE FACE: DECLARING CLIMATE EMERGENCY MEANINGFULLY

In the same statement to the Commons, responding to a request to declare a 'climate emergency', Perry said, 'I don't know what that would entail'. This statement can be read as an invitation, which XR can answer:

Our interpretation of declaring 'climate emergency' is that it is something that needs declaring by government, in direct address to the British people, and needs filling out in such a way as to make it meaningful. A whole-Britain-vision is needed, a war-time-style mobilisation. No one minister, however forward-looking or open-minded, can own this declaration. (Thus we ask for the declaration to be brought to Cabinet.)

The government should declare a climate emergency that requires quarterly COBRA-style meetings of the prime minister and key Cabinet ministers to track progress towards net zero emissions. Furthermore, the targets in the Climate Change Act need to be made annual or at most biennial, not five-yearly. And the Treasury Green Book *needs to be made green*, with this and more in mind. (We also need to talk about the biodiversity emergency; about the heart-rending and deeply worrying fact that we have wiped out about 50% of wildlife in the last fifty years.)

Telling the truth about the climate and biodiversity emergencies changes everything – it changes how we approach health, national security, emergency-planning, infrastructure and transport, water. The most crucial example of all, though, is food. Climate emergency means the very real risk that, before long, we will not have enough food to sustain ourselves.

TELLING THE TRUTH ABOUT LOOMING FOOD INSECURITY

The climate emergency that we want the government to acknowledge means many things. But what it means most urgently for the UK is potential food emergency. As temperatures rise and weather becomes chaotic, our crops (may) get baked and broken. This is urgent, existential as a threat for us in this country, because we are one of the countries in the world most lacking in food sovereignty. Unlike some countries, we do not face a dire threat of water shortage in the coming years of climate chaos. *But we are likely to face real food insecurity.*

Everyone can see that the weather is getting more and more weird. In the UK we had loads of sunshine this past summer, which many citizens experienced as a pleasure. But those heatwaves also hurt our harvests and pushed up prices. Unlike Spain, for example, Britain's agriculture is nearly all rain-fed and open to the elements (as is most of northern Europe). Soil depletion, a topic on which Gove has been encouragingly vociferous, will profoundly affect our capacity to grow food. Our weather is set to get weirder and more extreme still. This is all very bad news for the farmers here and abroad who grow our food.

Anyone who has tended to a garden or an allotment knows that even a brief spell of hot or cold weather at the wrong time can ruin seedlings. This is what we're now seeing globally as our weather gets more and more extreme. It will take only an annual sequence of individual extreme weather events in Europe, India, China, and North America to disrupt the global supply of wheat, maize, and rice. Our children need those calories. Will we be able to feed them three times a day?

The UK government isn't preparing Britain for the weird weather that's already disrupting our food supply. What if Europe experiences a few more hot summers in a row? Will we be able to import the extra we need from Asia and North America? Are we left to pray that the extreme weather leaves other parts of the world alone? If there's the slightest risk we won't be able to feed our families three meals a day, then the government must wake up, come clean, and act now. They should be protecting our children from the threat of hunger by *ensuring* the country can feed itself despite the impact of extreme weather here and abroad. Emergency investment in extreme weather-resistant food production is needed, plus preparation for unprecedented drought; appropriate crop diversification; drip-irrigation, and more.

And, of course, we need courageous cuts to carbon emissions, and an urgent summit on saving what's left of the Arctic ice-cap, so we have a chance to calm down the crazy weather.

There is a lot of information out now on the impacts of the 2018 weather on northern European agriculture, averaging 30% declines in output across grains and vegetables. Key is that this was across the northern hemisphere, not just in one or two countries, raising the potential of 'multi-bread-basket failure'.

The reasons are clear: lower temperature differentials between the Arctic and further south slows the jet stream, makes it wavy, and messes with the Arctic vortex, which splits into two, the two processes meaning we see extremes of weather. It is inevitable we will see more of this, as seventeen of the hottest years on record have been in the last eighteen years. We are seeing the slowest regrowth of ice ever recorded for this time of year, so the Arctic will be even warmer from now on; and thus more wild weather in Europe is coming. The coming years will often be like 2018 or more weird.

In the UK we have four months of global grain reserves. Our global economy can't take a hit of a reduction of supply of more than 30% for more than a couple of years before we have less grains-based foods in our mouths than we need for basic feeding. In addition, even before then prices will take grain-based foods out of the mouths of the poor.

To sum up: The fundamental fact about Britain in this context is that we cannot feed ourselves, *even now*. It is not credible to assume that we will be able to import vast quantities of food forever in a world experiencing increasing climate shocks.

The government needs to lead on ensuring that we have food for the future. Government needs to tell the truth about this threat – and start an urgent programme of climate adaptation. The country will not go along with such a programme with the speed required *unless there is honesty about the threat.*

CHAPTER 13

TRUTH AND ITS CONSEQUENCES: A MEMO TO FELLOW REBELS ON SMART STRATEGY [37]

I wrote this pamphlet as an oblique response to Roger Hallam's splendid pamphlet-book 'Common Sense for the 21st Century'. My own pamphlet confesses my massively pleasant surprise that the April 2019 Rebellion succeeded as well as it did, but argues that we cannot expect the same success if we simply repeat the same actions. We have to change up, and target those elements of the system that must change in order for there to be overall change of a sufficient magnitude.[38] This pamphlet enjoyed a largely positive response from the movement: see, e.g., chapters 16 and 17.

This condensed pamphlet seeks to tell the unvarnished truth about Extinction Rebellion past, present, and future, and, in doing so, to confront some difficult 'home truths' about the task that lies ahead. If we confront those truths, we may yet succeed in our wildest and grandest dreams.

We will soon be at the stage of our struggle where the focus shifts from seeking agreement from government, media, business, and

37 My colleagues in the XR Political Circle have kindly helped in discussing this document with me and helping to disseminate it, but its content is my responsibility alone; some of my colleagues don't agree with some points in it! I'm grateful to Sarah Kingdom Nicolls for heavy editing of this pamphlet, and to Adam Woodhall for some genius storytelling tips.

38 Further detailed support for this claim can be found in Section 2 of the Appendix to this book.

citizens that there is a climate and ecological emergency, to realising that if the emergency is going to be meaningfully tackled on the scale and timescale that we are calling for, relatively rapid changes in everyone's lives and institutions must take place. When that penny drops, we in XR are going to have to be even smarter in our tactics and strategy. By then, support for us from the wider public is no longer going to be 'cost-free' for them. A key focus of this pamphlet is suggesting what navigating that shift must mean for us. A key claim I make is that our struggle is different from and in fact much harder than those we have been inspired by – the Indian independence movement, the civil rights movement, etc. – because we have to change the entire system. A key implication I draw is that we need to tell and show the general public (by our actions, by what we 'target') that the chief burden of that change will fall upon those with huge unearned incomes: the rich and powerful elite.

TELLING THE WHOLE TRUTH

It's time the whole truth is told.

XR's first demand is for government to tell the truth about the gravity of the climate and ecological emergency. Part of how XR has succeeded so far is by way of us helping to break the climate silence, daring to be truthful about the imminence of potential societal collapse as a result of ecological collapse, and being willing to be emotional and not just factual about this.

We must be aware, however, of the potential perception that if our demands are met, then everything will be OK. No. As earlier chapters of this book have made clear, everything is not going to be OK. Our climate seems already to be spiralling out of our reach and the sixth extinction crisis is well underway. Climate disasters are coming, inevitably; and the climate situation will worsen for at least a generation, probably far longer, whatever we do, because of the time-lags built into the climate system.

We have to be honest about this grim truth. While we need to focus on prevention/mitigation (on our zero carbon target), our focus is even more critically required on adaptation: both transformative and deep. Truly understanding the scope of the climate and ecological emergency confronting us requires us to think and act more deeply than we ever have before. We need to be big enough to face climate reality in toto, in this way.

Don't get me wrong: I love Demand 2. In particular, XR's demand for the UK to go carbon-neutral by 2025 is an 'impossible demand'. It is completely politically 'unrealistic'. That, to me, is its great virtue. For XR exists to make the 'politically unrealistic' realistic – through a massive shift of consciousness and of will. This demand evinces starkly the extreme gravity and rapidity of the societal change we have to make. But it would be a very, very brave activist who would bet on it being achieved! And what that implies is: we have to start taking seriously the need to try to adapt to the worsening ecological situation that we and our children are almost certainly going to inherit *even if we achieve most of our goals.*

So, our rebellion must be as much about trying to create the seeds for something better to come out of the likely wreckage of this civilisation, as it must be about one last desperate push to change this civilisation into something ecologically viable without suffering catastrophic collapse first. We will experience enormous benefits if we radically re-localise, rebuild community, become less hyper-mobile and more energy-descended (i.e., less energy dependent), and so on, but, tragically, we need to think as much about the chance of achieving that *after* a collapse as to *prevent* a collapse.

TAKING STOCK IN LATE SUMMER 2019: WHAT XR HAS ACHIEVED SO FAR

The global situation is desperate. This is not a series of 'protests'. This is a rebellion. We are in rebellion against a government that cannot be conceived of as legitimate while it is committing its citizens to an unprecedented mass-suicide. When the 2019 April Rebellion launched, very few people thought it would succeed in the ways it did. I certainly didn't (I *hoped* it would, but I didn't *believe* it would). When I went onto the streets of London on 15 April and saw how few of us there were, just a few thousand across the whole city, I thought: It's not enough, the media will be ferocious to us, and the police will be rid of us within three days.

Well, the media *were* ferocious to us – for those first few days. But as we managed to hang on (by the very skin of our teeth, on Waterloo Bridge, and through extraordinary determination in waves of rebels at Parliament Square and Oxford Circus), as the police struggled to master the situation (or sometimes seemed not really to have their hearts fully in the job), day after day, our message started to get out there. The public mood started to shift. Swiftly.

We saw great determined willingness on the part of those arrested to make that sacrifice, and great emotional resonance – great truth-telling and story-telling – from arrestables and from spokespeople, and from our remarkable media team. Disruption helps makes the phenomenon real: it gets talked about and it becomes urgent to resolve. At a deeper level, people feel involved by the disruption. They feel as if they matter. The feeling of annoyance at being disrupted then has the opportunity to evolve into a feeling of respect for the disruptors, or at least of respect for the sacrifice and heartfelt message.

This is how we won in April,[39] combined with some luck, good weather (including a couple of unseasonably hot days that made our points for us), good timing of the Attenborough *Climate Change: The Facts* documentary, a widespread desire to have a chance to talk for once about something other than Brexit, and of course Greta Thunberg's visit to London and her declaration of support for us.

39 For more detail, see https://morningstaronline.co.uk/article/f/we-could-get-20000-or-30000-people-willing-take-direct-action-streets

On the Wednesday of the first week of the April Rebellion, we were still being pilloried by the media; by the Wednesday of the second week, an avalanche of change had come, including such extraordinary developments as a letter in *The Times* from business leaders supporting XR, and a major *Telegraph* op-ed by William Hague. A week later the XR Political Strategy team was having useful meetings with the Mayor of London, the Shadow Chancellor, and the Environment Secretary; and, incredibly, a motion declaring a Climate and Environment Emergency passed the House of Commons unopposed. Meanwhile, opinion polls showed, unprecedentedly, a massive majority of the public believing there is a climate emergency, declaring that they would vote differently because of that emergency, and 'the environment' shooting up the public's agenda. Proof positive of our success.

Furthermore, incredibly, Theresa May legislated for a carbon-net-zero target. There was simply no chance that this would have happened as recently as Easter 2019; the Rebellion made politically thinkable what before had seemed 'extreme'. Sure, the government's 2050 target is way too late, the government is still not telling the truth about our responsibility for the carbon-deadly GHG emissions that Britain causes through air and sea travel and through consumption emissions (aka embodied emissions), the scheme includes awful carbon offsets, and it contains a get-out clause if other countries do not legislate similarly. But it is nonetheless a historic and substantial advance.

Again incredibly, six Parliamentary Select Committees have united to create a kind of Citizens' Assembly to look at the crisis. Sure, the assembly they are creating has no legislative power, will meet for too short a time, and is unlikely to be asked about hitting carbon net zero by 2025; but it is still a historic and substantial result. *We did this.* Mass civil disobedience and nonviolent direct action work.

We basically won the first round of this struggle, XR 1.0.

In the next stage, we can expect greater numbers to gather in support and potentially to join us in nonviolent direct action. But we can also expect things to get harder. Because, as I've already noted, when attention starts to really shift from our Demand 1 to our Demand 2 – when people start to take seriously that serious action could be taken as a result of our NVDAs – then public support for our goals will no longer be a free lunch. This pamphlet focuses henceforth on negotiating that tough transition.

ACTIVE HOPE

So, let's tell the truth. We need government to declare an emergency, and to follow up that declaration with a massive truthful public information campaign *and* then to act accordingly: e.g., starting with moving swiftly to ban coal mining, ban fracking, slap duties on imported carbon, stop all airport expansion, and so forth. Demand 2 is obviously unattainable if we continue to move in the wrong direction as a country.

Demands 1 and 2 would require the prime minister to address the British people, acknowledging that things will still continue to get worse for a long time to come, that a long Blitz spirit and 'wartime mobilisation' will be needed to cope with this, and that massive resources need to be devoted to adaptation, as insurance in case of failure. Specifically, it would mean fessing up in increasing detail to the likelihood of food and water shortages and dire weather disasters in years to come. Britain cannot feed itself; our food position is chronically unwise and precarious.

Only once all this honesty is flooding our society can we realistically hope to actually triumph in the implementation of our demands.

OUR COMING LEAP INTO THE UNKNOWN

The great success of our April Rebellion might incline one to think that we can simply carry on following the tried and tested pattern of how past mass nonviolent revolutions succeeded. But one of the main burdens of this pamphlet is to suggest that that would be a serious mistake.

The task XR is engaged in is truly historically unique, unprecedented in scale and timescale.

Unlike the suffragettes, the civil rights movement, and the Indian independence movement, this is not a struggle to enfranchise excluded people. Women and black people could be accommodated into the existing system; in this way, the task of the suffragettes and of the civil rights movement, while hard, was do-able. But we want – need – to rapidly change the entire economic, social, and political system; within years, not decades.

Social science can tell us relatively little from now on. We in XR like to cite Erica Chenoweth's research about how NVDA can bring about regime change with the active support of only 3.5% of populations. But it should be borne firmly in mind that this has never occurred in a Western industrial 'democracy', let alone with the issues of climate and ecology taking centre stage.

BRINGING PEOPLE WITH US

We want our three demands to become law. Such transformation will mean that many economic interests get challenged or indeed ended. When it starts to impact people's lives for more than a few days' worth of disruption – when it affects jobs, what foods and products people are able to buy, or the ease of long journeys – then we will have to go to a whole new level in order to win the argument.

The first hard truth that we need to tell, in order to be in a position to win the argument that massive rapid change is needed, is – as set out at the start of this chapter – being clear that we cannot guarantee against collapse, even if our objectives are achieved. Our best hope for winning people over lies in authentically relating to them (and, crucially, to Citizens' Assemblies – see below) just how much of an emergency this really is.

But we cannot rely on even the full truth about the crisis doing the trick for us unless it includes the differentiated responsibilities of poor and rich. *Gross inequality must be tackled.* The stats are very clear: it is overwhelmingly the rich who are driving us over the ecological and climate cliffs. This is on an international/global scale, too. There simply is no remaining carbon budget for the 'development' of the 'developing' world, so rich countries must lead from the front, whilst learning – especially from indigenous and peasant peoples – how to return to life and to humility, by rapidly reversing our ecological footprint; by 'de-growing'. (This is the real meaning now of 'climate justice': contraction and convergence.) Of course, in the global context we Britons *are* the elite, in terms of carbon use, etc., and we are pretty much all complicit. It is only right that XR started here, for we started the Industrial Revolution and thus the course of events that have led to this crisis. But my case is that we have already provided that balancing. That is to say: we have already made very clear, in the actions we took in the April Rebellion, that we are not letting ordinary people off the hook.

What we need to do now, I submit, is re-balance. We must also not be perceived as hitting ordinary or poor people hard, or we will spawn a 'gilet jaunes'-style response. The French 'yellow jackets' hit the streets because they felt unfairly targeted by Macron's blunt regressive carbon tax proposal. We will be perceived as part of an 'elite' unless we direct more of our power against the elites. We cannot expect the ordinary person on the street to change if the rich aren't going to change.

Let's design the October Rebellion to bring that point home. We need by our actions to convey to the public that we are on their side; that, while everyone will have to change, the greater burden of the changes will fall upon those more able to bear that burden; that we are them (the people, the citizens, the 99%) and they are us. The XR 'theory of change' is incomplete if it does not include this. If these truths are told, then the populace at large may be willing for the kind of shared sacrifice that is needed. This has crucial implications for our strategy and tactics.

There is also an important point around sequencing here. If you think that we are already in a fully revolutionary moment (e.g., that a significant enough part of the populace – millions – is willing and ready to rise up and rebel), and that we can win completely very soon, then it makes sense to go for broke now. But if you think that we still have a hell of a long way to go to achieve Demand 1, so that the emergency really lands with people, then it makes sense not to hit ordinary people again. I want to make the claim that the October Rebellion ought to be focused primarily against the elite: the rich and powerful, especially those with vested interests in the fossil fuel industry, including crucially big finance.

THE OCTOBER REBELLION

The police will almost certainly be tougher on us in October. They will have learnt from last time and they will be anxious to avoid the criticism that fell on them last time (particularly from the 'Right' of the political spectrum). We must continue to seek to win them over, in the ways that were, it seems, remarkably successful in April. But we must be ready for harder responses from them, too. Goaded by the media and politicians, they may well clamp down on us in new ways, if we give them excuses for doing so. Furthermore, they will have had the chance to plan, and they now know much more clearly how our tactics play out.

Next time, we cannot risk near-dire mistakes (of which there were several near-misses in April[40]): anything we do or even threaten may be used as an excuse to pillory us, and maybe even to round us up on conspiracy charges.

It is imperative that our actions remain respectful. We should empathise with the difficult role assigned to the police; let politicians who believe their hands are tied by voters (and funders) understand that Citizens' Assemblies can help; and be clear it is the abysmal financial system (not the individuals staffing it) that is the key factor in the escalating ecological destruction, a system which must be replaced with something

40 I have in mind, for example, the 'plan' during April to bring the Tube to a standstill.

massively fairer and more intrinsically caring of humans and the ecosystem. We should target that system.

We need a laser-like focus on the kinds of actions most likely to generate active support and wider sympathy. We need overall to strike a balance between nonviolent disruption that is noticed by all and resolutely nonviolent disruption that falls, justly, more heavily on the rich and powerful. Above all, we need to rely more on our creative power, and on our philosophy. We need to generate a whole new level of emotional resonance.

TARGETING AIRPORTS

I am pleased to note that the XR Heathrow drones action proposed for September has been called off. (An action will probably go ahead along similar lines, under a separate 'Heathrow Pause' banner.[41]) My sense was that before we do airport actions, we needed people to understand that this is about whether there is food on the table in the years to come, or not. They mostly don't understand that yet.

If we do target an airport, it should be London City Airport, which is disproportionately used by the wealthier and business elites.

'TARGETS' IN THE OCTOBER REBELLION

Focus partly on parliament/government itself. By occupying Parliament Square in April, we made an impact. How much more powerful our action will be if we actually bring some of the business of government to a halt through NVDA – for days, or even weeks. For example, if we were

41 It did indeed do so, separate from XR: https://rebellion.earth/2019/08/02/statement-from-extinction-rebellion-uk-on-the-heathrow-pause-action/ Though the lines were dangerously blurred, with Roger Hallam's close identification with the Heathrow Pause actions. Heathrow Pause came to nothing: https://news.sky.com/story/heathrow-drone-protesters-blocked-by-signal-jamming-as-two-arrested-11808171 & https://www.theguardian.com/environment/2019/sep/12/heathrow-third-runway-activists-arrested-before-drone-protest .

to make it very difficult for parliament to enact a Budget that wasn't a climate and ecological emergency Budget; or if we were to massively disrupt a Ministry (e.g., the Treasury) from functioning at a time when it was seeking to enact dubious new rules or subsidies. Of course, a potential danger of such an approach is that a government starts acting more dictatorially in response. So we need to focus on *economic* elites too, to put elite material interests in the balance, and to make clear that reducing democracy would not stop us from being effective:

Focus on major targets in the city. Imagine shutting down the Stock Exchange – actually stopping it from operating for a significant period of time, stopping its profit-engine-of-destruction. Or imagine shutting down Goldman Sachs, or indeed a bunch of merchant banks. Imagine, perhaps, occupying and shutting the Green Investment Bank, which the government privatised and neutered, or Canary Wharf, which is very geographically vulnerable due to limited access roads. Doing any of those would be hugely popular with a citizenry still seriously pissed off by how the 'banksters' got away with murder financially in, and ever since, 2007–8. And it would concentrate the minds of the elites greatly. Power has never conceded anything without a powerful demand that threatens it.

We can also see parliament's concessions as being made to neutralise us: seeming to concede towards our demands. If the state refuses to move beyond gestures, blocking real action into 2020, then there must be serious consequences. We need to concentrate the minds of power and the authorities further. If we do this, we potentially bring the citizens onside in such a way as to ready them to make some changes in their own way of life, too. For those changes need to be broadly, democratically agreed.

CITIZENS' ASSEMBLIES: HOW TO MAKE THEM REAL

We want Citizens' Assemblies (CAs) that have real decision-making power, to decide how to act sufficiently to rise to the challenge of the unprecedented long emergencies we face. Governments are likely to try to fob us off with purely advisory CAs – as the Select Committees' initiative already does.

The plural is crucial. We want Citizens' Assemblies, not just for the four nations of the UK, but for localities across those nations, too. Local government just as much as national government needs to be revived and democratised. They need to be sortitionally based, a part of democracy, much like juries already are.

CAs will be tasked with figuring out, with expert advice, how to put together the drastic package of changes, the as-wartime mobilisation, now needed. But why would politicians give away some quasi-legislative power to the CA? CAs can take issues out of the 'too-difficult box' into a zone where something real – enough – can be done. Politicians can offload onto the public, who are picked to deliberate in the CA the difficult responsibility of acting for the benefit of the future.

One possible way of enabling politicians not to lose more control than they are willing to, is to include some of them in the CA. This was done in Ireland in the Constitutional Convention several years ago and helped move forward public discourse and action in relation to such vexed matters as gay rights, abortion – and climate policy.

It turned out that including politicians in a powerful CA, far from (as some had feared) leading it either to inertia/gridlock or to domination by those elected politicians (who were a third of the Constitutional Convention), helped legitimise it in the eyes of the Irish Parliament, and helped enable some of its recommendations to come to pass. The option of agreeing to a CA with elected politicians among its members is therefore one with a powerful positive precedent, and it is an option that XR should actively retain.

However, once again, let's remember that we are in an unprecedented situation here: public consciousness is changing with extraordinary rapidity. Politicians smart enough to recognise this as both an ecological truth and increasingly a *political* one may already be willing to be considerably bolder than Ireland was.

To win a CA with real power from the October Rebellion or thereafter, we need to strike a balance between:

1. giving politicians the sense that we can help them out of a hole;
2. giving everyone the sense that everything really does have to change, and that it's time to get serious about making that change;
3. giving everyone the sense that in that process, some – roughly, the 1% – have to change a lot more than others.

The key take-away is this: We need, by our actions, to give fellow citizens the sense that the CA will be just and seek justice. For even if a Citizens' Assembly were to come up with radical enough plans that we'd be able to lay down our Rebellion for good, we cannot expect those plans to be implemented if they appear to the majority of citizens to be unfair or hostile to their interests. But when people understand that the rich are the ones who need to cut back the most and that this is going to happen, then we become the popular ones, and, crucially, we pre-empt a 'populist' reaction against us.

THE CRUCIAL ROLE OF OUR CHILDREN

There is one more critically important element to the picture, without which success in October remains improbable. Namely, our children. If the power of youth is added to the power of truth, then that equals potential transformation of the kind required. The climate school strikers have called for a general (adult) strike – the Global Strike – on Friday 20/27 September. It is absolutely crucial that we show massive solidarity with this, and help it work. It is critical because our children and their plaintive calls to 'Save our world!' are our most powerful instance of emotional resonance; their struggle has touched the world.

Moreover, Greta Thunberg is (in my opinion) a world-historical figure on a par with Martin Luther King or Gandhi.

We are the ones above all who must and will heed the kids' brave call. And then they may hear our call. Imagine an October with tens of thousands of children and students on the streets of London engaging in NVDA. Rebelling. That is how a crucial stage in the civil rights movement was won, in Birmingham Alabama – with over a thousand minors in the prison cells. It was a step not taken at all lightly by the civil rights leadership. But it worked. It could work here too.

Our October Rebellion follows swiftly on the heels of the 20/27 September marches. It must seek to reach out to involve people of all ages, especially those children. When there are arrests and imprisonments of children alongside adults, the game will have changed: the authorities will be placed in a very difficult dilemma; whatever action they take will look very bad. If we can forge a close alliance with the school climate strikers this October, then the unprecedented historic change I've been outlining in this pamphlet could come to pass. Perhaps humanity's darkest hour really will become its – our – finest hour.

ON BRAVERY

If we can look climate reality in the eye, if we can bear to face the extinction crisis that we have engendered, and if we can respond to these with open-ended flexibility, adaptability, and courage, then perhaps a new hope arises. Like many XR colleagues, I am confident that our ultimate purpose is not just political but psycho-spiritual. Our deepest purpose is to manifest the spirit of transformation, whether or not we succeed in our goals. Our rebellion is about doing the right thing: being dignified and courageous in the face of adversity, no matter what the consequences. The trick is not to be attached to *getting* the right outcome, while working determinedly and as intelligently as possible *for* the right outcome.

We open ourselves with courage and caring bravery that lies buried deep within our collective body to see and feel the true terrible state of this beautiful world, how badly we have damaged it and especially how badly we've damaged our other-than-human animal kin and our descendants. When we really do this, there is no alternative but to rebel. In this sense, our ecological perception, our psycho-spiritual awakening and our socio-political uprising are all different aspects of the same whole, the same process.

I believe history will judge us very kindly. Either we succeed, we are 'glorious failures' (paving the way for something better after us), or we fail completely, in which case there won't *be* very many more history books…

Our Rebellion is a cry from the heart, a here-I-stand-I-can-do-no-other, as much as it is a calculated effort to try to achieve a particular set of very bold and (in this late hour) almost impossible policy outcomes.

All we can do is our best, our smartest, and our soulest. We may succeed in rebelling against collapse and possible extinction. There can be no greater prize. It is in this sense a great privilege, albeit also a dark and painful one, to be alive at this fateful moment in history.

Let's keep our eyes on that prize.

Rebel for life.

CHAPTER 14

RUPERT READ INTERVIEW ON BBC RADIO 4 'TODAY' WITH JOHN HUMPHRYS: ON A REPORT BY POLICY EXCHANGE ATTACKING XR

When the so-called 'think tank'[42] Policy Exchange produced a report viciously attacking XR and personally pillorying Roger Hallam and Gail Bradbrook, Radio 4 decided to lead with it the following morning. It was decided I would go up against veteran Today *programme attack-dog John Humphrys. I had virtually no sleep that night but was ready to do nonviolent battle early the next morning. In the interview that ensued,[43] I took the opportunity to question whether Humphrys was doing his job as a journalist by not challenging Policy Exchange, especially over their funding. Sure enough, in the following weeks,[44] it emerged that Policy Exchange had taken* funding from fossil fuel interests.[45]

---

42 See https://www.thetimes.co.uk/article/big-companies-buy-influence-with-funding-for-think-tanks-6x85mpx9q , https://powerbase.info/index.php/Policy_Exchange , https://politicalscrapbook.net/2013/09/policy-exchange-pay-us-140-to-watch-our-youtube-channel/ , https://www.theguardian.com/commentisfree/2017/may/17/dark-money-democracy-billionaires-funding and the references below.

43 https://rupertread.net/media-appearances/dr-rupert-read-bbc-radio-4-today-discussing-extinction-rebellion-including & https://youtu.be/TIBxjdMECqg – you can listen to the interview here.

44 The interesting coverage that followed included 'In the Age of Extinction, Who is Extreme?' by Simon Mair and Julia Steinberger https://www.opendemocracy.net/en/oureconomy/age-extinction-who-extreme-response-policy-exchange/, 'Think Tank Won't Reveal Who Paid for Report Calling Extinction Rebellion Extremists' by Adam Ramsay: https://www.opendemocracy.net/en/dark-money-investigations/think-tank-wont-reveal-who-paid-for-report-calling-extinction-rebellion-extremists/ & https://rebellion.earth/2019/10/08/bbc-we-talked-about-this-so-why-are-you-still-letting-vested-interests-undermine-peaceful-movements/ .

45 As established by *Vice* magazine: 'Group That Called Extinction Rebellion "Extremist" is funded by Big Energy', https://www.vice.com/en_uk/article/ywagdx/policy-exchange-extinction-rebellion-funding .

John Humphrys: For eleven days last April many streets in the centre of London were blocked solid by supporters of the XR campaign protesting against climate change. Now they have taken their protest to other cities, including Cardiff, Glasgow, Bristol, Leeds, as well as London again. Here's Greta Thunberg, the young climate activist, addressing protesters in London in April. [Greta Thunberg – speech to crowd.] On this programme an hour ago, the former head of counter-terrorism at New Scotland Yard, Richard Walton, described XR as a hard-core anarchist group who want to break up our democracy. [Excerpt: Richard Walton: I was asked by Policy Exchange to have a look at XR and I thought it would be just like any other environmentalist group. Sadly, after a lot of deep research, it's very clear that they're a hard-core anarchist group that want to, basically, break up our democracy.]

JH: [Stern and serious] Well, Rupert Read of XR is with me, good morning to you.

RR: Good morning.

JH: What do you make of that?

RR: Well, I think it's a funny kind of revolution that's trying to break up the state and break up democracy if it takes a year for a top counter-terrorism officer to dredge up a few quotes that he can try to paint a funny picture around in order to make the case. What we *are* is a mass movement of ordinary people of all ages and all kinds who are coming together to stand up because we're facing an existential threat, we're facing the possible end of our civilisation and people aren't happy about that, and so yeah, they want radical change, we want radical change; don't we all?

JH: Well, we want change, a lot of people want change, certainly, but that depends how it's brought about. So let me quote to you what one of your leading people, Roger Hallam, said in February of this year: 'We're not just sending out emails and asking you for donations, we're going to force the government to act and if they don't we will bring them down and create a democracy fit for purpose. And, yes, some might die in the process.'

RR: Well, I'd rather live on a healthy planet than die. But if we look at what happened with the suffragettes, if we look at the civil rights movement, people did die in those movements, and this movement, this struggle, is even bigger than those. But, do you know what? If people die it's not going to be because of us, because we are an absolutely nonviolent movement. The nonviolent discipline that we showed in April, which has been absolutely strong, is one of the key reasons why the British public swung behind us. Because that's the fact, that 67% of people, after the rebellion in April concluded, agreed with us, that there is a climate emergency and the government needs to act now.

JH: [Loudly] But! That's one thing, it's a big step further beyond that to say 'we are going to force the government to act, and if they don't we'll bring them down and create a democracy fit for purpose. And, yes, some might die in the process'. That's an enormous step beyond what you just said. Isn't it?

RR: Well, Mr Walton said we want to break up democracy –

JH: [Interjects loudly] I'm asking you to comment on what Mr Hallam, your Roger Hallam said.

RR: Look, if you go through the enormous, voluminous works of Roger Hallam, I'm sure you can dredge up one or two quotes that I might disagree with the nuances of.

JH: Ah, so you do disagree with it?

RR: Well, I wouldn't necessarily have used those precise words myself, but what I would like to bring out from those words that I do strongly agree with, is that what this is not about us breaking up democracy, this is about, as Roger was saying in that quote, *creating* a real democracy, because everybody knows…

JH: [Urgent interjection again] Who, whose democracy is that?

RR: [continues through interjection] … that democracy is failing us.

JH: Sorry, whose democracy is the real democracy?

RR: The real democracy is the democracy we want to create. If you'd asked Gandhi what he thought of British democracy he would, I'm sure, tell you that he thinks it would be a good idea. How can it be democratic for us to be…

JH: [interjecting] Because there was no democracy as we understand it when Gandhi staged his protests [chuckles derisively] because it was a colonial country. You're not surely making a comparison with this country? We have a vote, in this country, it is a democracy, and you want to bring it down?

RR: Look, 67% of people said, 'Hey there's a climate emergency and the government needs to act now to deal with it', but that's absolutely not happening, and that's why we'll be back on the streets until it does happen.

JH: Right!

RR: That's why our rebellion in October will be stronger and bigger than the rebellion in April, and if anyone listening to this wants to come and join us, put October 7 in your diary, that's the date it's going to start, and we will be back on the streets until the government acts [JH attempts to break in] and faces up to this terrible crisis [JH interjection attempt] which is going to destroy us…

JH: And doing what? What else apart from protesting in the street, what else are you prepared to do?

RR: Well, we'll be taking action probably against financial institutions, we'll be taking action…

JH: [interjects] What's that mean, 'taking-action against them'. What's that mean?

RR: Well, it might mean things such as blockading them, it might mean things such as having some kind of debt strike, there are…

JH: [interjects] Debt strike?

RR: ...various kind of ways.

JH: Can you explain?

RR: Yes. It might be refusing to repay debts that are taken out from financial institutions that are acting completely irresponsibly because they're putting our very future at risk, and that's what this is all about, let's keep coming back to that, John, because this report, this report is trying to take aim at XR and saying, 'Oh look what Roger Hallam said about this or that.' This report seems completely uninterested in science, completely uninterested in children, completely uninterested in our future, and *that's* what this is about. This report, the only purpose of this report is to defend business as usual, and it's business as usual that is killing us, and what I'd love you to do, John, is to ask who is funding this report; who funds Policy Exchange? Because that's what you didn't do when you talked to Mr Walton earlier. The funders of Policy Exchange are completely non-transparent. But I bet you, if we were to find out who they were, we would find out in whose interest it is to undermine XR. For example, maybe it's the fossil fuel companies.

JH: That may be, I've no idea, but Richard Walton...

RR: [interjecting] Well, why don't you find out, John? Why didn't you ask Mr Walton about that?

JH: Because I was talking to him as former head of counter-terrorism at New Scotland Yard, which is what his job was...

RR: [interjects] But who funds him, who is his paymaster *now*? That's the real issue. Because we deserve to know about that. [JH fails to interject several times] This isn't about the words of Roger Hallam or Gail Bradbrook, this is about who's trying to undermine Extinction Rebellion, and I think many of your listeners will agree with me that actually what we ought to be doing is getting behind a movement that is trying to save our future, not throwing barbs at it.

JH: A movement that is prepared to break the law.

RR: Absolutely, we're prepared to break the law, we're a rebellion, the clue's in the name, because how can a government be fully legitimate that is putting us on a path, a serene path, if you see what I mean, to destruction, they're serenely taking us down the road to mass destruction.

JH: [interjects] And you want to, and you want to bring down capitalism. That's what your tweet in April said. 'This movement is the best chance we have of bringing down capitalism.' I mean, you are aware that capitalism has produced some of the wealth in this country that has enabled people to have a decent life and you are aware of what capitalism does?

RR: Look John, I think that everybody knows that our current system is failing and that's what we're asking: We're asking people to stop and take a look at our system and be ready to reassess it, because if we don't do that and fast then we're in dire trouble, and António Guterres, the Secretary General of the UN, says we've got not ten years but two years to start seriously acting if we're to get this crisis back in some kind of tenable state, and that was last year that he said that, so we've got one year left to start acting seriously! That's why we'll be back on the streets, that's why we're willing to break the law, yes, that's why eleven hundred people got arrested and transformed consciousness around this issue, in this country, for the better. And that's what's giving us some real hope for the future.

JH: And just one final thought then. You want our GDP to stop rising. 'Drop the rising GDP; it is nothing to be proud of. Before or after we cause our civilisation to collapse.' So, in other words, you want there to be a permanent state of recession in this country. That is your position, is it?

RR: Look, on the one hand we've got business as usual, which is what this report Mr Walton is asking for, is what Policy Exchange is asking for...

JH: [speaking over Rupert] Right, so recession, that's what you're arguing for. It's a straightforward question.

RR: The alternatives, John, the alternatives are, on the one hand business-as-usual and climate break-down and mass death, on the other hand we reassess and try to find a way of going forward together so that we can actually live and prosper together, and I really don't think that's too much to ask. I really don't think that's remotely extreme. That's common sense.

JH: Rupert Read, many thanks.

CHAPTER 15

BRITAIN'S DEMOCRATIC STRUCTURES ARE BROKEN: CITIZENS' ASSEMBLIES COULD FIX THEM

I wrote this piece to frame a potential 'XR-ish' response to the prorogation of parliament (later ruled illegal by the Supreme Court).[46] I had hoped that XR might seize the moment to help the British people to understand that Extinction Rebellion is not just about the climate and ecological emergency. It is also a call for a deepening of democracy – for real democracy; *for the people (the demos) to govern (kratein) – as the only way to deal with political problems too big to deal with any other way. To me, this piece highlights a road sadly not travelled.*

Boris Johnson's planned suspension of parliament has highlighted just how fragile and unfit for purpose Britain's 'democratic' structures are. The decision to prorogue parliament for five weeks – the longest prorogation since 1945 – is a transparent attempt to block parliamentary scrutiny of the Conservative leadership's Brexit plans or lack thereof. The fact that some of their media defenders have tried to paint this as in any way normal is absurd. This is an abuse of a parliamentary instrument that by convention has been used only for short breaks preceding

46 https://medium.com/@GreenRupertRead/britains-democratic-structures-are-broken-citizens-assemblies-could-fix-them-and-more-61b91bb33e09

a speech from the throne. Politicians, the public, and much of the commentariat are rightly outraged by this cynical exploit.

But while this outrage is not misdirected, we should not be lulled into seeing this simply as an aberration instigated by an aspiring tin-pot demi-despot within an otherwise functional democratic system. The reality is that our so-called democratic system is riddled with flaws that undermine the legitimacy of our parliamentary representative democracy. Boris Johnson's prorogation has simply stretched that legitimacy even further.

One of the most obvious ways that our democracy is broken is the fact that most votes simply do not matter in our first-past-the-post (FPTP) electoral system. That is how it can be that 11% of the UK electorate in 2017 voted for the Liberal Democrats, Green Party, and UKIP combined while between them these parties only got 2% of the seats in parliament. (The vote for these parties would, of course, likely have been way higher were they not subject to the dubious – but under FPTP understandable – 'wasted vote' argument.) The fact is that minority political viewpoints are almost entirely without any parliamentary representation in this country. This leaves the very small percentage of the population who are members of the Labour or Conservative parties in a vastly more enfranchised position than the rest of us because they get additional votes over who their leader or MP candidates are going to be. This is a disaster at a moment when we desperately need a more fluid, agile responsiveness in our political system to the needs of the day and to the wishes of the electorate.

There is a bitter irony to the number of commentators and politicians who virulently support or are conspicuously silent about the disenfranchisement of millions of voters through the FPTP system but are infuriated by Boris Johnson misusing prorogation in this way. Indeed, to defend the first but decry the second significantly weakens the moral force of the outrage. Both are, after all, perfectly 'legal' ways of shutting down diverse parliamentary viewpoints.

As well as these legal disenfranchisements, we have the almost blanket deprivation of prisoners' voting rights in this country, and the legal but possibly immoral denial of the franchise to sixteen- and seventeen-year-olds. We also have the erosion of local democracy by central government, which threatens virtually to make the position of Councillor redundant. And we have the vexatious question of how we go about representing the fundamentally unrepresentable future generations who will inherit climate and ecological breakdown.

Clearly, we have deep questions as a society about what sort of democracy we would like to create. And while I have ideas about what sorts of policies would widen democracy, I am humble enough to admit that my own personal democratic manifesto is not going to solve the deep lack of democracy in our society. Instead, I think we should all be taking a leaf out of the handbook of demands put together by the Extinction Rebellion movement and looking towards Citizens' Assemblies to solve some of these issues of democracy.

Politicians cannot be at the heart of fixing a political system that they themselves are beneficiaries of. That is why so many Labour politicians still support the FPTP voting system; it keeps their representation vastly over-inflated in parliament. It is why Conservative politicians oppose extending voting rights to sixteen- and seventeen-year-olds; young people not voting keeps Conservative MP numbers up. The fact is both these parties have historically been and continue to be the beneficiaries of a non-democratic system.

These are not issues that parties with vested interests should be trusted to lead on. Instead, we should truly return control back to the people by creating a series of Citizens' Assemblies to discuss, debate, and ultimately craft a constitution to fix this broken democracy. These assemblies would consist of randomly selected members of the public by sortition in much the same way that juries are selected. The selection process would also seek to balance demographics so that no region, gender, ethnic group, religion, or sexuality is underrepresented.

This model was used successfully in Ireland to draft recommendations in response to a whole range of social and environmental issues.

Additionally, we should make space in Citizens' Assemblies on those other issues aside from constitutional reform that parliament has shown itself unable to grapple with. There are at least two such massive issues today: most pertinently we have the climate and ecological crises that, if we do not get under control, threaten to collapse our society. XR has made a compelling case for a Citizens' Assembly/ies to chart the way towards climate- and eco-safety by 2025.

Added to this is, of course, the vexing issue of Brexit itself, which politics-as-usual has palpably failed to settle.

I submit that Citizens' Assemblies could well provide illumination on all these issues and extend democracy outside of the grasp of our zombie-like two-party duopoly.

We should seize the opportunity that Boris Johnson has given us through his constitutional vandalism by highlighting the deeply broken nature of our democracy and use it to call for drastic reform of our system. Perhaps then something good may yet come of what at first glance seems to be another nail in the coffin of Britain's claim to be a founding bastion of democracy.

Three Citizens' Assemblies: one on democratic and constitutional reform, one on the climate and ecological emergency, one on Brexit. This is how the UK could heal itself.

CHAPTER 16

HOW A MOVEMENT OF MOVEMENTS CAN WIN: TAKING XR TO THE NEXT LEVEL

My pamphlet 'Truth and Its Consequences' (Chapter 13), achieved significant resonance. Responding to my central claim there that XR's October Rebellion should target primarily the rich and powerful, Rob Hopkins, co-founder of the Transition Towns movement, said: 'Excellent. Amen to that'.

This piece[47] aimed to rebut some misunderstandings of the pamphlet, and so to make clear how consonant my intended approach was with XR principles and values.

Our October Rebellion needs to be bigger than April's was. Way bigger. It has to be about not just getting verbal concessions from power, pious declarations of climate and ecological emergency without actual consequence. No; that's just not good enough. Because this is the age of consequences and there's no more time to play with.

No less than António Guterres, UN Secretary General, has said that if we don't start making serious – transformative – change within

47 https://medium.com/@GreenRupertRead/how-a-movement-of-movements-can-win-cfcfdad5151c

about the next twelve months, there is no way the world can head off the kind of catastrophe outlined in the IPCC's 2019 report 'Global Warming of 1.5°C'. The Budget this autumn is the last chance that the UK Government has to alter its spending priorities hugely within this time-frame. That Budget must be a Climate and Ecological Emergency Budget.

If the authorities won't budge, then there will be consequences. Us. Forcing a fork in the road. From our October Rebellion, we need a massive change of direction conceded and acted upon. And if not, then a Citizens' Assembly to decide the way forward. An assembly of ordinary citizens, properly informed about the hell-on-Earth that humanity is creating, and empowered, after real reflection and deliberation, to decide on how this country is going to play its part in charting a very different path. Fast.

How are we going to get to this new scale of rebellion? How do we get several times as many people out onto the streets this autumn, and many more of us ready to be arrested, and some of us ready to be imprisoned?

Anyone who was there in April will tell you it was the experience of a lifetime. I believe that pretty much every one of us who was there in April will be back in October. Plus those who regretted not having been a part of it; they'll be there too. And many more who have been 'woken up' by the talks, articles, and TV interviews they've heard since.

But there's no guarantee that that will add up to enough new rebels. How do we go to the next level?

We have a plan: a movement of movements. We're seeking contingents of new rebels from movements 'allied' to XR: the peace movement, the animal rights movement, the social justice movement. And several more. Call it the Rebel Alliance...

My pamphlet 'Truth and its Consequences' (Chapter 13), proposed a stratagem that could underscore this potential rebel alliance, explaining how we could and should target the economic and financial system and those who run it. The power behind the authorities. This can bring together our movement with others that are devoted to targeting particular pillars of the system. Those most deeply concerned about social justice, or about the distorting effects of money upon our society and polity.

Let me dispel some possible misconceptions about this proposed approach to our strategy for this autumn.

This is not about targeting rich and powerful individuals. It is therefore not about blaming, naming, or shaming. This is about targeting the *system* they run, and sometimes targeting their role in running that system. It is about system change.

XR is beyond party politics and not a competitor on the standard political spectrum. So this is not about being 'left-wing'. No, the logic is much simpler than that. In an emergency, we need to be all in it together. But that simply means that it's elementary that the polluter elite need to (for example) reduce their flying more than the rest of us, and so on and so forth. Flying needs to be rationed, or at least to be done on the basis of a stringent progressive 'frequent flyer levy'. There was nothing 'left-wing' about instituting food rationing in World War II; it was simply what made sense in the spirit of all pulling together. The same kind of spirit is needed now: a spirit of wartime mobilisation in this greatest of emergencies in human history. Similarly, there's nothing 'left-wing' about Contraction and Convergence, the cleanest and clearest proposed way of achieving some basic level of climate justice, converging on us all polluting the same small amount that the planetary ecosystem can now tolerate, worldwide.

This is about making very clear that the systems that are running us over a cliff-edge are intolerable, and that they must change. Now. We will not allow the City of London to continue to leach money that should be used for an emergency programme of transitioning our economy to something that can be ecologically viable. We will not allow them to continue to invest in fossilised energy-systems that are killing our children, right now.

I hope, given the relatively decentralised way that the October Rebellion will be run, that you'll choose, alongside others, to put your body on the line in stopping parliament from instituting any Budget that is not a Climate and Ecological Emergency Budget, or in stopping the financial 'heart' of London from beating its deadly drum of business-as-usual. I hope you'll be part of the Rebel Alliance that is poised to change this country forever, and put it on a fairer path; a path to a future.

And after they fight us, and we 'fight back' with the force of truth and in deepest nonviolence...I hope *we'll win*.

CHAPTER 17

STOP THE SHIP SINKING

The October Rebellion in 2019, was the second long UK/international rebellion, an NVDA uprising. In my pamphlet 'Truth and its Consequences' (published a couple of months prior to the rebellion, and reproduced here as Chapter 13), I had called for us to target London City Airport and the City of London itself, rather than just blockading streets. The October Rebellion saw us doing those things, and they were some of the most successful parts of the rebellion, which made me pleased and hopeful. This article, which was published on Medium *halfway through the October Rebellion,*[48] *focused partly on those aspects of it.*

As of 7 October 2019 the biggest people's rebellion for climate justice the UK has ever seen recommenced and I, amongst thousands of others, am immensely proud of the turnout in the seven days so far. With all hands on deck, about twice as many as we had in April, people from all over the country are driving the movement for the huge wave of change this society requires.

So, now we are entering our second week of the October Rebellion, and we are targeting the City of London as well as Westminster. And so I am calling everybody to come down to either London or, if that is impossible, to their nearest demonstration, and experience how it feels to support a better future for generations to come. It is the presence

48 https://link.medium.com/O744nOgTK5

of those who advocate the truth of our climate and ecological emergency that creates the electric atmosphere we have felt around central London over the past week.

This also comes from the sense of duty bestowed on us by other international movements, such as Fridays for Future, led by teenagers such as Greta Thunberg.

Fridays for Future (FforF) is our children rising up to beg for the right to live. It is shameful for our whole civilisation that our children have been driven to this. As I see it, Extinction Rebellion is the adults rising up to heed the children's call. Our children are calling on us all to provide for them a future – if you really heed that call in its full depth and desperation, then I believe that you will join XR in the streets. Now.

XR and FforF are powerful allies. Greta Thunberg came and supported our launch in October 2018; I had the honour of eulogising her after the speech there. At a meeting organised by *The Guardian* while she was in London during last April's Rebellion, I asked Greta directly whether she supported XR. Her reply was very simple and equally direct: 'Yes, I do.'

This greater movement of our movements acting together in concert is supported by the many of us who simply want to be on the right side of history. Some police officers who accompany us on the streets have been widely reported to support our goals. Not surprising really: the police have kids too. Moreover, in this new phase of rebellion we have even been joined by former police officers, willing to risk arrest by taking part.

It is the nonviolence of XR that is the secret to our success. The police can handle violent protest easily; they have a well-established protocol to follow, and it 'legitimates' any heavy-handedness on their part in response. Mass nonviolent rebellion, such as ours, is far, far harder to deal with. We are thousands of ordinary people putting our bodies on the line for a just cause. That makes it very hard for the police and the authorities to stop us.

If anyone ever acts violently, they are not XR. Our rebellion is one of peace. We rebel because of the love we feel for the natural world, for each other, for our children. We demonstrate what the future can hold, having learned from the best examples of progressive and courageous movements in our history. Like predecessor movements such as Otpor in Serbia, People Power in the Philippines, the suffragists, Martin Luther King in the US civil rights movement, and Gandhi in India, XR is, in its nonviolent discipline and determination, showing exactly how a civil resistance should look – and our purpose is clear.

XR is for the enriching of democracy. This is why our third demand centres upon the need for Citizens' Assemblies. We want citizens involved in deciding how to change everything: to succeed where politicians have failed to adequately address the climate and ecological emergency. This is, of course, the reason that some politicians make unpleasant statements about XR. It is precisely because they realise that we are exposing them and their failure for all to see. It isn't easy for politicians to countenance the breaking of the law, because they make the laws. But everyone knows that sometimes the breaking of laws is exactly what is needed: when those laws are themselves wrong.

That is why it was justified for Martin Luther King and the civil rights movement to break the law, to quote a famous example. Everyone now regards them as obvious heroes – even though, at the time, what they did was very controversial and they attracted much hatred. Those breaking the law with peaceful direct action or skipping school this week – for the sake of changing the law and changing the system so that humanity doesn't destroy itself – will without any doubt be regarded as heroes later. They – we – will all be pardoned. The best thing for all decent politicians to do would be to act now so that we don't have to keep breaking the law.

XR is about everyone coming together to struggle for a common future. The struggle is like a war, but not against each other. It's a completely nonviolent struggle where the enemy is our own history, our own extant infrastructure, our own complacency. XR, unlike some previous radical movements, welcomes everybody into a common effort for the very survival of our species.

The ship of our society is sinking. It will without doubt succumb unless we do something extraordinary and fast. XR is our last best hope. I urge you to come and see what we have started. The October Rebellion lasts for just one more week, during which we need the support of every conscientious member of society. We must address policy makers, firmly demanding of them to ACT NOW.

I look forward to seeing you on the streets.

CHAPTER 18

CANNING TOWN TUBE INCIDENT: AN APOLOGY

Within XR, this might well be the most controversial item in this book. This chapter reproduces the text of a comment I posted on my personal Facebook page on 18 October 2019 in response to an incident at the Canning Town tube station. What happened was this:

In the second week of our October 2019 rebellion, the police sought to make all XR protests in London, even legal protests, illegal. This produced an enormous backlash; suddenly XR had many NGOs and civil liberties organisations on its side. The momentum was with us. But a tiny affinity group had plans to try to disrupt the whole tube system, based on an earlier plan XR had decided not to carry out. This threatened to derail the momentum that XR was by then enjoying and many of us sought to dissuade this tiny group from going ahead. The planned action, disrupting public transport in a potentially dangerous space, was very unpopular within XR, let alone outside it (a last-minute poll conducted within the organisation showed overwhelming opposition). Sadly, the action went ahead, with consequences even worse than most of us had feared. Violence ensued, including a defensive kick from one of our activists as he was hauled off the top of a tube (and then beaten up).

XR's national Political Strategy group decided unanimously to ask the movement to pause and to atone for this serious mistake. But this didn't happen; the national media team continued to pump out material praising the tube action even though it was an unprecedented public-relations disaster for us. As more and more people asked me to speak out against the action I felt torn; I certainly didn't want to undermine the brave well-intentioned rebels who had undertaken it, but I didn't feel I could remain completely silent, either.

The privacy-setting on my Facebook post was marked as 'Friends', but someone cut-and-pasted it onto Twitter. At that point I thought it pointless or wrong to back away from it; I went with it. The response to my apology was overwhelmingly positive, from both inside and outside the movement, but for a minority of rebels what I said was considered a kind of treachery, so the matter remains a controversial one.[49]

If I were writing the statement now, I would word it differently. It was written in the heat of the moment, in pain and deep concern. It deserves to be reprinted unchanged.

I deeply regret that the action on the tube went ahead this morning. XR Political Strategy group, to which I belong, advised strongly and unanimously against it, as did the vast majority of the movement.

Lessons must be learnt so that never again can the actions of a tiny number of 'XR' activists tarnish the entire movement. Once it was clear that this action was going ahead, XR should have disowned it, yesterday. But we didn't (it seems) have a process for doing so. In future, the process for such disownment, where necessary, needs to be clear.

The point of the action was worthy: to demonstrate the utter frailty of the tube. If climate chaos is not reined in, the tube will flood repeatedly and then terminally, and be taken from us forever. A small amount of disruption now might help prevent a vast disruption to come. (Imagine what a nightmare London will be, if and when the underground network floods.) But the design of the action was questionable, and its

49 I've since met with several of the tube action rebels themselves, most of whom now agree that the action was a strategic error, and I enjoy very good relations with them. 'We are all crew', as the XR saying goes.

execution obviously flawed. One of those on the train at Canning kicked out in self-defence at those who were threatening to injure and possibly lynch him. Understandable, but not nonviolent – XR is always nonviolent. The timing, after yesterday's wonderful gathering at Trafalgar Square to protest the government's authoritarianism, during the hugely significant court case we are bringing against that authoritarian withdrawal of the right to lawfully assemble across London,[50] was quite simply catastrophically stupid.

I'm just grateful to those commuters who stopped the protesters who were being beaten up from being really badly injured. Their spirit is what has impressed me most from this whole sorry saga, and leaves one with some hope that the human future may yet not be doomed.

(FYI: The Political Strategy group is calling for a pause in XR's October Rebellion, while we reflect on and atone for this most difficult moment in the movement's history thus far. Watch this space.)

50 Brilliantly, this court case was subsequently won, rendering null and void hundreds of the arrests during October, and exposing the police force's serious breaching of reasonable response.

CHAPTER 19

AN INTERVIEW WITH PROFESSOR BENJAMIN RICHARDSON: LOOKING BACK OVER THE FIRST YEAR OF XR

This interview is forthcoming in a special edition of the Journal of Human Rights and the Environment *on 'From Student Strikes to the Extinction Rebellion: New Protest Movements Shaping our Future'.*

1) Why was there a need to create an 'Extinction Rebellion' (XR) given the existence of many competent activist environmental organisations campaigning for urgent action on climate change, biodiversity loss, and other threats to the biosphere? What are the key objectives of XR?

Extinction Rebellion's objective is to instigate radical action in response to the climate and ecological crises. We believe that this requires systemic structural change across all levels of society. We direct our campaigning towards pressuring governments into adopting radical ecological policies, while also seeking to engage in consciousness raising among the general public and through the media.

I first became involved in Extinction Rebellion in early September 2018 when the movement was in embryonic form. I came to it with the background of someone who had invested a lot of work into pursuing action on the climate crisis, primarily through the route of party politics. I had previously been a Green Party councillor, a national Green

Party spokesperson, and I had run for election to parliament and the European Parliament multiple times.

While I still support this route to effecting change, I became increasingly anxious that it was not delivering tangible results fast enough. The way that most people speak about climate within the public sphere involves a severe obfuscation and downplaying of just how threatening the crisis is. This collective denial makes it hard for radical political parties to build enough public support to have their policies implemented. We need grassroots organisations like Extinction Rebellion to change the narrative about this crisis if we are to have any hope of rising to meet it.

Extinction Rebellion has succeeded where so many other institutions have failed in drawing attention to this crisis and building support for tougher action among the public, media, and even politicians. The work of big green NGOs and political parties has simply failed to deliver results fast enough. Extinction Rebellion, combined with the school strikes for climate (and in particular the advocacy of Greta Thunberg – who is XR's most important supporter of all), has succeeded in making the vital inroads that we so desperately need to get started on climate action. (This is clear in the UK from polling data, which makes clear the impact of XR's April 2019 Rebellion upon climate consciousness in the UK.[51])

2) To what do you attribute the success of XR in mobilising support?

Part of Extinction Rebellion's success stems from its decentralised structure. Local Extinction Rebellion chapters throughout the world have been empowered to set their own campaigning agendas and strategy providing they stick to our ethos of radical ecological action and nonviolence. This has empowered people to become involved in their local groups.

Our strategy of nonviolent civil disobedience has also proven extremely effective in generating media coverage and provoking political discussion. When compared to some of the law-abiding marches for climate in the

51 See, for example https://www.carbonbrief.org/guest-post-rolls-reveal-surge-in-concern-in-uk-about-climate-change

past, Extinction Rebellion has proven that we can generate a lot more discussion about climate with far fewer resources, if we are willing to risk arrest – but remain completely disciplined in our nonviolence.

Our way of speaking – emotional, authentic, truthful – has also been very important. This is illustrated by our colleague Greta Thunberg's stinging, visceral speeches excoriating world leaders for their 'betrayal' of young people by failing to tackle the climate crisis. In my own media interviews, such as with the BBC on 9 October 2019,[52] I've argued passionately that Extinction Rebellion is essentially about 'whether or not our children have a future'; on that occasion, I showed to the cameras live on TV the names of my two young nieces tattooed on my forearm, a temporary-tattoo I'd had done to remind me of what this is all about, and to show to the police if I was being arrested.

Finally, Extinction Rebellion has also found itself as a focal point for action in a society that is beginning to wake up to the existential threat of climate breakdown that threatens its own demise. As our climate deteriorates and more people become alert to that fact, Extinction Rebellion is the organisation best placed to channel resistance to climate and ecological breakdown.

3) To what extent is XR co-ordinated globally, and by what means? Or is the movement largely self-directed and shaped by local groups acting under the rubric Extinction Rebellion?

Different chapters exist in different countries, but things are decentralised rather than centrally co-ordinated. Subscription to the ideologies of radical ecological action and nonviolence are shared universally among all chapters of Extinction Rebellion.

4) Does the XR have any affiliation, formal or informal, with other environmental/social organisations and/or movements, such as the wave of school student-initiated climate strikes around the world in recent months? If so, what is the nature of that collaboration or connection?

52 You can watch the BBC 'Politics Live' show here: https://www.youtube.com/watch?v=0FNSVIfWTqs

There is a lot of overlap in agenda between Extinction Rebellion and the School Strikes for Climate. These strikes embody the same ethos of nonviolent civil disobedience, self-organisation, and radical action on climate and ecology that Extinction Rebellion has. Consequently, there is overlap in the two movements' goals, strategies, and even the campaigners involved in both. Though we think it is very important that the School Strikers retain their full independence; the children are morally leading our society now.

In relation to other organisations, Extinction Rebellion also shares many of the ideals of ecological NGOs and even of some politicians. We aim to work constructively with people from different groups, while maintaining our independence and ideological integrity. As such, we have held meetings with senior politicians from across the political spectrum and sought to impress upon them our demands for tough action on climate and ecology.

5) Does XR have a presence outside Western countries, and in parts of the global South such as India, China, or Brazil, and if so, how is that presence evolving?

Yes, there are national chapters in many countries across the world. However, given the decentralised nature of Extinction Rebellion, I am not best placed to relay all the work that they do. But I can tell you that XR UK has given these chapters much support, including financially.

6) What are the principal tactics of XR, why those tactics, and how do they link to the overall strategy of XR?

A key tactic of our organisation has been mass nonviolent civil disobedience, as mentioned earlier. We have had over one thousand people arrested in our April 2019 Rebellion and nearly two thousand in our October 2019 Rebellion. This has helped generate a lot of media attention about our cause, and even forced the issue of declaring a climate emergency into parliamentary discussion.

7) What evidence is there that mass civil disobedience is effective in leveraging cultural, political, and/or legal changes on environmental/social issues?

The declaration of a Climate and Environment Emergency by the UK Parliament directly after we met with them following our April Rebellion is a prime example of our strategy yielding tangible successes. Another is the amount of media coverage we were able to generate and keep in the news cycle over a prolonged period during our April and October rebellions. There are also opinion polls which directly suggest that our April action in particular led to an increase in the general public's concern about the climate crisis: at the end of the April Rebellion, two thirds of Britons agreed that there is a climate emergency.

8) In regard to XR's invocation of a climate/planetary 'emergency', is there a risk that this strategy will reinforce the promiscuous vernacular of emergency in our political culture – e.g., Trump's declaration of an 'illegal immigration' emergency and earlier rhetoric about a 'drugs use' emergency? More specifically, might the declaration of an emergency imply a suspension of the rule of law, thereby allowing draconian measures to be introduced that ultimately create additional problems for keeping governments in check? So, what do you see as the positives and risks of the language of 'emergency' for XR's aspirations?

Emergency talk is essential. It is truthful. As I explained in an essay published in *The Guardian* some years ago: 'I do not agree that we should leave aside talk of "catastrophe". In fact, by sticking to talking of "climate change" rather than of "climate chaos" and "potential climate catastrophe", we end up playing the same game as the more subtle and intelligent of the climate change deniers by adopting their language'.[53] And as climate communication campaigner Jane Morton has argued, emergency talk can move people to action when they are confronted with emotionally engaging depictions of serious

53 Rupert Read, 'Emergency Talk,' *The Guardian* 13 November 2007, www.theguardian.com/commentisfree/2007/nov/13/emergencytalk

threats that are personally relevant, while offering pathways for finding solutions.[54] This is an emergency if anything ever was.[55] It is a long – basically permanent – emergency. It will define all our lifetimes – increasingly. We guard against the negative risks of emergency talk by our approach being based in love. This is a 'compassionate revolution'.

9) What have been some of the key lessons that XR groups have learnt over the past year of activism and organising? What are the internal processes in the movement for reflection on tactics and strategy and drawing lessons from these? How can XR sustain itself, given the risk that it might eventually lose novelty and media interest, and thereby lose momentum?

Our regenerative culture is key here. XR has been designed in such a way as to minimise the risk of burnout. Although it rarely gets the same level of media coverage as XR's other activities, the regenerative culture is deeply woven into our mission. The first element of this regenerative culture is to shift the accusatory narrative from individual behaviour towards shared, empathetic feelings of grief about ecological losses and our children's futures. Secondly, XR is devoted to an ethic of care – eschewing violence and promoting the well-being of XR activists, especially those subject to arrest or other adversity. Thirdly, XR has an ethos of collaboration and co-operation – a stance known as 'we are all crew'. XR actions are built not on personal, ad hoc interventions but on acting together and building relationships in the process. Finally, XR's regenerative culture is about 'relinquishing the ego', by eschewing attention on celebrities and leaders in order to create space for the voices of many, especially those less privileged.[56]

54 Jane Morton, 'Why We Should Mention the Emergency', https://climateemergencydeclaration.org/janemorton

55 However, it is also true that it is not widely perceived as an emergency, even despite XR's extraordinary agenda-changing achievements in 2019. That is why we wrote 'Rushing the Emergency, Rushing the Rebellion?', collected as the Appendix to this book.

56 See Anna Pigott, 'Extinction Rebellion's "Regenerative Culture" Could be Just as Revolutionary as its Demands', 2 May 2019, www.opendemocracy.net/en/opendemocracyuk/extinction-rebellions-regenerative-culture-could-be-just-as-revolutionary-as-its-demands

10) In light of its experience so far, can we expect to see any new directions in XR, in terms of its goals and tactics over the next few years? And if the political aims of the movement are not substantially addressed within the next couple of years – before critical planetary boundaries are breached – how might XR react?

A new direction for XR will likely be to go beyond targeting governments to challenging the private sector, especially major corporate polluters and the financial sector that bankrolls them. Pressure must be stepped up against fossil fuel companies and their investors, such as pension funds, as the global divestment movement does, through external disruption (NVDA) and 'internal disruption' (speaking truth to power). Such disruption can help stigmatise and undermine these businesses' social licence to operate. If there is not more action in response to our movement within the next few months, then expect a greater emphasis on 'adaptation'. This will be us making clear that time is virtually up for mitigation-centric approaches.[57]

11) XR has been criticised[58] for failing to account for how structural inequalities, including racism, sexism, and classism, have shaped the climate crisis in its analysis. How has XR engaged with and learnt from these critiques? How can we discuss the existential threat of extinction that climate change presents to humans as a species whilst reflecting the complex realities of everyone's lives in this narrative?

Extinction Rebellion was set up with a clear emphasis on the democratisation of our governmental response to the climate crisis. We think that Citizens' Assemblies must be given a crucial and binding role in setting our route towards net zero emissions by 2025. A key motivation for including this demand for the expansion of democracy is the belief the Citizens' Assemblies are likely to be far more sensitive to creating an equitable and just transition – climate justice – than politicians. Clearly, we must make sure that the radical changes we need in our society are conducted in a way that reduces rather than exacerbates existing injustices.

57 See Chapter 27.

58 See letter from the grassroots collective The Wretched of the Earth: https://www.redpepper.org.uk/an-open-letter-to-extinction-rebellion/?fbclid=IwAR2Nq1xdQ8tQiuBp8_jgr_qrZBx-q2MEu3zwkXynMGGy2n5sTtf51f-Hx_yA

As an organisation Extinction Rebellion grew very fast. Consequently, we have not always been sufficiently clear in communicating demands about the importance of justice to ecological transition. As the organisation matures, we have sought to engage with the criticisms you have mentioned to make sure we are more fully integrating communication about justice in our messaging.

However, let me add that some of these criticisms are quite wrong-headed. It is an insult to those XR leaders who are people of colour to say that XR is 'white'. It is an insult to the working-class spokespeople and leading activists in XR to say that XR is 'middle-class'. (We expect such rubbish from the right-wing press; it is a shame when it comes at us also from our supposed friends.)

And the basis in a problematic version of 'identity politics' of some of these criticisms undermines them, for they fail to understand that ours is a compassionate revolution, that does not name, shame, or blame – whereas sometimes they do. And ours is a universalistic movement, broad-based – unlike the silo thinking of too much 'identity politics', which appears to be based in a politics of resentment.

12) XR has been criticised for how it has related to police and law enforcement and the assumption that the police will become sympathetic to XR.[59] In a context where we are seeing moves in Australian jurisdictions to increase police power and create new criminal offence in response to XR protests, can you reflect on the relationships between XR, protestors, and the police and how this might be changing?

Our attitude towards the police has been based in the pragmatism of being able to achieve greater success if one has a better working relationship with them. This worked well in April 2019. It didn't work so well in October 2019 when policing of XR actions in the UK became more hostile. We recognise that the police are agents of governments, which in a variety of countries, including the UK and Australia, are arming police with greater powers to thwart XR protesters. As I have

59 See http://criticallegalthinking.com/2019/04/29/extinction-rebellion-credit-criticism-cops/

explained in an essay I wrote with Dario Kenner, 'There is a balance to be achieved between continuing not to dehumanise the police – and indeed to appeal to them – whilst also making it clear they are obliged to defend the system that XR wants to transform – and thus not naively expecting them to 'defect' to us, and not making it look as though we "love" what they are and do'.[60]

60 Rupert Read and Dario Kenner, 'XR UK: Telling the Truth through Targeted Disruption', 29 November 2019, www.opendemocracy.net/en/opendemocracyuk/xr-uk-telling-truth-through-targeted-disruption

CHAPTER 20

2025 NO MORE? IMPLICATIONS OF THE CONSERVATIVE VICTORY FOR XR, AND FOR EVERYONE

The UK went to the polls on 12 December 2019 in what was supposed to be 'the climate election' but in fact was dominated by Brexit. XR mounted what we called an #electionrebellion, seeking through various forms of nonviolent direct actions (including hunger strikes) to bring public attention heavily to the eco-emergency as an election issue, but this did not appear to have a significant impact on the election campaign. The Conservatives won their biggest majority since the heyday of Margaret Thatcher. Prime Minister Johnson failed to turn up for the televised Channel 4 debate on climate and nature; Channel 4 replaced him with a melting ice-scuplture. But it didn't affect the result. This piece, co-authored with Frank Scavelli and published in Open Democracy,[61] *was my response, arguing that XR's demands require modification to include the increased likelihood, as a consequence of the election results, of eco-driven societal collapse.*

> 'In many countries the public knows the old promise of tomorrow being better than today is finished. But they don't quite know why that is, or what to do about it. ... We seem trapped within the dynamics and momentum of this system. Therefore, my guess is it means that we won't change things. At least not in time to prevent

61 https://www.opendemocracy.net/en/opendemocracyuk/2025-no-more-tory-victory-xr-and-coming-storms/

> catastrophe from a range of societal stressors, the most unavoidable of which is climate damage. So what to do about it? The first step is to stop pretending that we will prevent things from getting worse. Instead, to consider just how bad things will get and what in that context we could do to help. Once you have let go of those old stories of progress, there is nothing negative in working for a lesser dystopia.'
>
> – *Jem Bendell*

The Madrid round of annual UN climate talks in December utterly failed – a perfect example of world leaders lacking seriousness about the existential threat we face. Australia continues to burn; a perfect example of how very serious our vulnerability is.

Here in the UK, the party with by far the worst rankings of its climate policies – so bad that its leader preferred to be represented by a melting piece of ice rather than defend Conservative policies in live televised debate[62] – has been handed electoral victory and a landslide eighty-seat majority.

In the face of our new, darker post-election reality, the foremost question must be: Is Bendell right? The decisive Conservative victory and the total lack of serious international actions means all those sympathetic to Extinction Rebellion must undertake a period of reflection; reassess some of our demands, hopes, and perhaps methods and probably make some significant adjustments.

There is absolutely no way that our Three Demands Bill, nor anything like it, will be willingly passed by this parliament. Painfully, we have to admit to ourselves that the 2025 deadline of net-zero carbon emissions and biodiversity loss, perhaps our best known of the three demands, has suddenly become almost unachievable (barring a very risky and improbable seizure of power from the current government). For we are simply running out of time.

---

62 See this Euronews article about the Channel 4 News climate debate: https://www.euronews.com/2019/11/28/boris-johnson-replaced-by-melting-ice-after-ducking-election-climate-debate

The UK's contribution to the climate and extinction crises is not small, as the government misleadingly suggests it is. A significant chunk of China's seeming contribution to the problem is actually ours. We've merely offshored our emissions along with our manufacturing.

Even if the government will not meet Extinction Rebellion's first demand – to 'tell the truth' – doing so ourselves is our foundational principle. If the 2025 deadline is based on hard science and the precautionary principle – which it is – we cannot simply move the emissions deadline back to 2030. So our three demands should remain targeted firmly at 2025. But we must begin to accept and discuss what the future will look like as that 2025 deadline slips into unachievability.

The UK, 'standing alone' outside the EU, is primed for the mother of all falls. For, unable as we are to feed ourselves, we are deeply vulnerable to the vicissitudes of a deteriorating climatic situation for which we have much historical responsibility.

We might, through further, larger, smarter rebellion combined with attempts to awaken the governing elite, influence this government in a better direction, by influencing the consciousness of the many and hitting the profits of the few. We should of course try. But the reality we now face probably makes it too late to stop this civilisation from ending during the next generation or two. We must start thinking seriously and realistically about the possibility of society being likely to decline or even collapse. The situation also reflects our tragically broken democratic processes: a dire corporate media, a free-for-all zone of paid lies on Facebook, and a laughable electoral system. XR's call for a profound renewal and deepening of democracy is more pertinent than ever and will be very hard to implement under this government.

So yes, we must keep fighting against the self-destructive tendency of our society every step of the way, and keep pressure on government to reduce emissions and biodiversity loss drastically. But the dismal character of recent events mean we *also* need to start preparing ourselves and our society for the probability that our efforts will fail, and that we will undergo, within the next generation or two, some kind of collapse event. In that light, we must now start to take seriously the role of transformative adaptation and, yes, even what Bendell has termed Deep Adaptation.

We need to start talking about 'solutions' for Britain. We can no longer put this off, awaiting a Citizens' Assembly, which a Johnson administration won't make happen. We need to emphasise things like restoring wetlands, making nuclear waste and nuclear power plants safe against the coming storms (literal and metaphorical), reducing food waste, eating lower on the food chain, learning and sharing food-growing and other skills, and a thorough relocalisation. We need to be talking about all this, bringing to life the story of the climate and ecological decline that the Conservative manifesto will likely lock in. And we need to start thinking, fast, about how to *do* these things, including helping to bring about Citizens' Assemblies to help chart the way forward, with or without central government backing.

Above all, we must remain truthful. XR is most advanced in the UK, but even here we could not make real a 'climate election'. This painful fact is what requires the reassessment I have called for in this piece. If even in the UK, XR couldn't midwife an election result less climatically and ecologically inauspicious than this one, then we need to be more grimly realistic about the future and about adapting to it.

As climate and ecological deterioration become a fact of daily life, we increasingly need to rely on each other and our local communities. We must also therefore redouble our efforts to affect *local* governments and force them to take seriously the task of preparing our communities. We need to look more seriously at deep alliances with the Transition Towns movement, with permaculture, and so on.

This civilisation – growthist industrial capitalism – is in its endgame. Heartbreakingly, under the Conservatives, that process is very likely to accelerate. We in XR need to start being even more direct about this, and we need to make that directness real by starting to call for and to seek to enact transformative and deep adaptations. We are set free from fantasies of progress and salvation. Instead, living in truth, let's regenerate our movement and start to get clear on what is now to be done. On what can now still be hoped for – and what we have to let go of.

CHAPTER 21

CHANGE OF THEORY? AN INTERNAL MEMO TO XR UK RE STRATEGY

This memo, previously unpublished, was co-authored with Dario Kenner of 'Why Green economy?', and circulated within the movement in January 2020. It sought, in the wake of the successes and failures of the October Rebellion, to orient the movement more towards concrete targets that expressed the system-change we need, and away from targets that were interfering with ordinary working people's lives.

Our highlighting of the vulnerability of supermarkets and supply-chains is now somewhat superfluous; the coronavirus crisis did that in spades. The point now is to ensure that society learns from that and changes, fast.

Whoever won the General Election, it was pretty much always going to be down to social movements to dramatically push forward the green transition. Here are some thoughts on where XR goes next.

SUMMARY

We either stay as an 'environmental' movement or we are part of catalysing something much bigger.

We can't go on directly disrupting the public in ways that don't make sense to them and still expect to win. We need to build the movement

– we need to choose actions that accrue public support, focusing our disruptive power more on the powerful vested interests that are making us ever more vulnerable to climate disasters. This is not junking the XR 'theory of change'; it is including an intermediate movement-building phase in it, so that we can actually generate the much bigger number of rebels we need.

XR's actions contributed to the unprecedented rise in environmental awareness reflected in opinion polls last year. This led to competition between political parties on green issues in the election. At this crucial moment when our very survival as a species may be on the line and the context we operate in is getting worse, it's time to seriously consider the different options available to us. It's not enough having 3.5% of the population behind us if a majority oppose us. If we are to get real action in the early 2020s, we are going to have to go down a different route.

To win we need to combine XR's disruptive power with wide popular support. Let yourself dare to imagine how powerful this combination could be.

SENSE-MAKING *WITH* THE PUBLIC

In our previous essay[63] we called for XR to focus disruption on powerful actors who are blocking change, like government, and on corporates – polluting companies and the banks that fund them – as well as most of the media. For any government to be able to act meaningfully we will need to counter heavy lobbying by the polluting companies, hedge funds, banks, etc. who want to continue profitable destructive business-as-usual. That's where XR's unique disruptive power can help tip the scales.

But we must garner wider popular support or the government will be able to continue to dismiss us as 'middle-class crusties'. XR Action Strategy Update for 2020 noted that in October 2019, 'Our tactics

63 https://rupertread.net/writings/2020/xr-uk-telling-truth-through-targeted-disruption

struggled to cut through increased opposition from the media, police and our Government'. One of the key reasons that happened is because our critics successfully presented the false division of 'protestors' vs 'commuters'. As long as this perceived division exists, XR, as part of a movement of movements, just won't receive enough public support for the huge systemic changes that are required. Our critics, who are hugely influential over sections of the population that we need on our side, like to criticise our tactics because they know that actually the majority of the UK population believes, understands, that there is a dire environmental crisis. Let's flip the criticism and show the public that we are listening, learning, and adapting.

What would all of this mean in practice? Let's apply the questions below to all actions/strategies to make sure we are picking the best targets. When we're deciding on targets to disrupt that seem obvious to us, we must ask ourselves: how will it look to the public? Imagine disruptive actions with specific targets are successful, what then? In the design and communications around our actions we need to be able to answer obvious questions from the public and media. By thinking through these answers we can better help the public to understand that the disruptions make sense and the results are viable. This makes it harder for us to be dismissed as out-of-touch crusties.

Overall, ask yourself whether the actions that are proposed will strengthen the movement or (as predictably occurred at Canning Town) weaken it. Here are a few rough-sketch examples to illustrate the approach and the complexity in getting popular support.

HIGHLIGHTING VULNERABILITY: SUPERMARKET SUPPLY-CHAINS AS A POSSIBLE FOCUS FOR XR ACTIONS

a) Will this potential action directly undermine the powerful vested interests blocking change?

Yes. For example, simultaneous actions targeting the companies' corporate HQs and relevant government departments. Yes, especially, if we take pains to try to selectively impact what we seek to disrupt: e.g., can we hit products that mainly the rich consume? Non-necessities? Goods that are imported unnecessarily (because we grow them here too)?

b) Will it be clear to the general public why we are doing this action?

Work would need to be done to make clear to the general public why we were doing this. That could be done with a little simple storytelling, e.g., our just-in-time system makes us immensely vulnerable; Britain is not self-sufficient in food; we are vulnerable *so that* we can have rampant consumer choice and rampant profiteering by corporates – is this really a good trade? With smart messaging, public support could potentially be rallied *more* easily than occurred in April 2019: if we temporarily reduce supermarket consumer choice, but force a debate about stopping our out-of-control food system from putting us at risk of serious food insecurity, then we have done a very good thing. We can highlight the probability of heatwaves of unprecedented strength and duration here and in other 'breadbaskets', as is virtually certain on a business-as-usual scenario.

c) Will it be clear to the public why this huge issue threatens them? Will they therefore be at least potentially sympathetic?

It will be easiest to make the case on those occasions when disruption happens anyway: e.g., when fires in Russia compromise our supply of wheat, as happened a few years back. Obviously, if we ourselves create an 'artificial crisis', say by blocking a 'just-in-time' supply depot, then that raises the stakes.

But the beauty of such actions focused on revealing our shared vulnerability is that, provided they are well contextualised, the threat is clear. These actions are about the threat facing us all; they bring to light our intense vulnerability to climate chaos; and yet they specifically target those who are responsible for that system fragility (and indeed are making money out of it!), rather than being a blunt instrument hitting Joe Public.

DISRUPTING NORTH SEA OIL AND GAS PRODUCTION

a) Will this potential action directly undermine the powerful vested interests blocking change?

Yes. Disruption would affect the fossil fuel companies who are successfully lobbying to extract every last drop out of the North Sea. There are obviously locations that would directly disrupt them and not the general public. XR Scotland has acted already on these targets in early January. To go beyond what Greenpeace and others have done in a similar vein in the past, the proposed actions would need to be mass disruptive actions. With XR involved that becomes possible, which is exactly why the oil industry fears us.

b) Will it be clear to the general public why we are doing this action?

Yes. There is an obvious direct link between the extraction of fossil fuels with climate change. It would make sense to the public if Scottish oil and gas production were targeted because Glasgow will host the UN climate change conference in November (COP26).

To gain popular support it must be clear that these actions are in solidarity with the 300,000 direct and indirect workers in this sector. The disruption would be to create pressure on their employers not to abandon them when there is a transition away from fossil fuels: i.e., it would create pressure for a just transition. Any actions to call for the end of fossil fuel subsidies would need to focus on tax breaks for production. It would not be popular to call for an end to consumption subsidies because this would increase household energy bills.

c) Will it be clear to the public why this huge issue threatens them? Will they therefore be at least potentially sympathetic?

This might yet be the Achilles heel of this proposal. Targeting the fossil fuel industry directly is hard to connect directly to people's own lives; it feels remote. Although it's about the cause of the problem, it doesn't *tell a story* about that problem. This aspect needs the most work, if the actions are to be successful. Some potential messages include: burning oil and gas contributes to dangerous climate change; XR is acting for the survival and security of the public; oil is raw material to make plastics filling up our oceans and ruining beaches around the UK; oil is what goes in vehicles, causing deadly air pollution in the UK.

ANNEX: OUR ARGUMENTS IN MORE DETAIL

Let's be proud of what has already been achieved. A start has been made by getting parliament to declare a (symbolic) Climate and Environment Emergency, a carbon net-zero target (albeit one set far too late) and a Citizens' Assembly (albeit one without any real power). Now XR needs to use its disruptive power in concentrated ways to actually get the change that matches what climate and ecological scientists say is desperately needed.

We have already argued[64] that we need to bring the attention of the public squarely onto those whose responsibility for our shared predicament, our awesome vulnerability, is the greatest. We've argued that XR has already made the point in spades that we are all in this together and that it is vital now to enable citizens to understand that it is wealthy interests who must change the most. That a just transition will mean that ordinary folk will not suffer from the kind of drastic action that is required. Sure, change will sometimes be difficult or painful, but we can show that we intend to ensure that most of the pain falls on those who can bear it, e.g., the polluter elite. Those who can afford it. (Note again, this is not because we are all 'Lefties'. XR is not! It is because there is no alternative.)

We need to build the movement hugely, if we are going to be able eventually to return to the streets and win. This has to be 'story-led' – our actions have to be led by a clear story of how society (and especially 'the 1%') needs to change, plus an inspiring vision of what, in

64 See again https://rupertread.net/writings/2020/xr-uk-telling-truth-through-targeted-disruption

outline, the new society can look like. A vision that we need to embody in regenerative actions.

Example 1: Pop-up allotments in spaces where it is not legal to have them. This would be guerrilla gardening with a political edge: we would seek to defend these 'allotments' and genuinely grow food there.

Example 2: Having lots of XR local groups ready to physically respond next time there is extreme weather; this happened, inspiringly, in the aftermath of the recent flooding in the Doncaster area of Yorkshire. This would show XR doing something practical and help to counter criticism of pure 'virtue signalling'.

Regenerative actions embody the kind of future society we are pre-figuring. They take us gently into the space of 'solutions': a space we cannot avoid starting to occupy if we accept, as we should, that the chance is low of getting government to deliver us a Citizens' Assembly that answers the call we have made in Demand 3.

Such actions will be more local in nature but are still part of a strategic/national/global story. A story relevant to our seeking change from governments, local and national, because, in the wake of the General Election result, action locally will quite surely henceforth be as important as action nationally. Local actions can in this sense still be strategic actions: they include things like tackling those businesses where you live that are standing in the way of the radical action so clearly needed.

HOW DO WE BREAK DOWN THE FALSE DIVISIONS BETWEEN 'PROTESTORS' AND 'COMMUTERS'?

After a successful beginning to the October Rebellion, lamentably the image our critics and much of the public were left with was that of the XR 'protestors' being attacked by commuters for blocking the tube at Canning Town. The 'commuters', both present and online, celebrated this as a victory against the 'protestors'. This cannot ever happen again.

We must seek instead an approach which makes sense to the public, which reinforces our message, and which inspires people to join us.

In contrast, the April 2019 phase of the rebellion was an incredible success, placing XR on the map and establishing a profile for the movement. We don't need to disrupt the public, at least not directly, anymore. We can disrupt 'with' the public when we are a true mass movement, with numbers that will swamp the police, but we can only do that if we become much more popular.

How do we do *that*? We don't do it by just repeating what we did in April and October over and over again. We do not do actions which are designed to stop ordinary people getting around, inciting criticism of our actions. So, let's take away their chance to reject XR because of its tactics. Let's give the public the best possible chance to support XR, to make it easy and common sense to support us.

Why is this so crucial? Unfortunately, the December 2019 election was not a climate election even though XR (in particular, through our brave hunger strikers[65]) and other movements tried to make it so. There isn't going to be another election in the short term so that route is not open to us. It might even be in a certain sense good for XR that Conservatives rather than Labour won; a Labour victory might have made the movement complacent; the Conservative Government are sure to act in ways that stoke the fire in our bellies. We are almost certainly going to have a government in place for years which has so far been very hostile towards us. The only way we are going to successfully pressure it, and to defeat the power of the polluting industries that lobby it, is if we have a broad cross-section of the UK public more or less on our side.

We make our Three Demands relevant to ordinary people so that they can see we are not just another group of 'environmentalists', or as some media frame us 'green zealots'. We need to connect to what is

65 https://www.independent.co.uk/news/uk/home-news/extinction-rebellion-hunger-strike-conservative-labour-party-climate-change-a9207886.html

already happening in people's lives. Especially those who are living precarious lives – for climate chaos is about to make all of us more precarious, but the already-precarious most of all. We need to do this so that people see the 'environmental' issues we are raising as the social issues that they are. Issues of everyday life, like transport, housing, and – above all – food. There were lots of XR actions making these links during the #electionrebellion. We need more of this, on a mass basis – connected to mass actions. And storied. That is, placed in a narrative that makes sense to people.

The starting point for XR actions from now on has to be that we reach and connect with people who do not consider themselves 'environmentalists'. We will not be able to convince everyone to join XR but we do need widespread sympathy and support. We can do this by making clear that what is at stake is not mainly 'environmental' concerns. It is concerns of survival and security. And by making clear that we are also intent, via our actions, on showing how a climate-sane society *will have to be* a more equal society, a society where the rich and not the poor bear the main burden of change.

To conclude. We are not saying don't take actions that will cause any disruption to ordinary people. We are saying only cause actions that disrupt ordinary people in ways that can be made sense of. And we are saying we must no longer (appear to) *target* ordinary people for disruption: instead, we must aim at those most responsible:

> the corporate media, including social media, which we've seen out of control recently in the extreme anti-democratic manipulation of Facebook ads. Make it possible for the truth to be told.

> the polluter elite. They must change if there is to be any common future.

> the fossil funders. Everyone knows they cannot go on like this.

> the giant supermarket chains. They are at the root of our everyday vulnerability.

CHAPTER 22

XR: FINANCE'S FRIEND?

This piece was commissioned by Reuters (Breaking Views), *which is read mainly by those in the finance 'industry'. They altered my headline and the opening paragraph in misleading ways, causing the unfortunate impression – which some in the movement were understandably unhappy with – that I was somehow seeking to ally XR with actually-existing* Haute Finance! *The version published here is my preferred original version.*

I wrote it while engaged in an endeavour to tell the truth starkly in Chatham House rules meetings with members of the elite and the Establishment. It is interesting how many of this elite are actually quite aware at some level that the system literally cannot be sustained. What they typically lack is a clear picture of what to do about it that is adequate to the crisis. I like to give them some pointers: e.g., They should make it explicit that it's OK for employees to join Extinction Rebellion actions.

Why does Britain matter in the world today? Because, let's be honest, it doesn't matter very much. We are an increasingly small country drifting noisily into the Atlantic. We are small fry.

But there are still a few ways in which we really matter. Here are the top three:

1. Our past matters: because the Industrial Revolution started here – which in its speed, scale, and mode of economic organisation has recklessly taken the world to the very precipice of

untrammelled catastrophe. This means that we have a greater historical responsibility to right the profound wrongs of mass habitat destruction and of climate genocide than any other nation on Earth.

2. Our present matters: because of the City of London. Yes, we're smaller than ever – but we punch way, way above our weight in terms of our species' impact on the natural world, in that our country finances that impact. Fifteen percent of all fossil finance passes through the City of London. That must change, if there is to be any future for life on Earth (including your life, reader).

3. Our future matters: because the great arising consciousness of how everything has to change is manifesting more strongly here in Britain than anywhere else in the world. Extinction Rebellion, the movement that has most profoundly shifted eco-awareness and that has made possible conversations at the highest levels of power both financial and political that could not have happened as recently as just a year ago, was born here and is burgeoning most strongly here. Britain 'led' the world into the terrifying mess it now finds itself in; Britain still 'leads' the world in financing that still-growing mess; and so, quite rightly, it is Britain that should be sniffing a way out of that mess...

WHAT IS XR'S ROLE IN THIS?

Under pressure from the persistence of our Rebellion last April, power moved to concede something to each of our three demands:

1. Parliament declared a Climate and Environment Emergency – albeit, only symbolically (*government* needs to declare such an emergency, for it to be legally binding);

2. Government legislated for carbon net zero – albeit by 2050, far too late;

3. Parliament has set up a Citizens' Climate Assembly – albeit one lacking hard power.

What's next for us? Given the December 2019 UK General Election result, bashing our heads against the levers of political power is not likely to be the most effective way for us to get further traction in 2020. So we are likely to turn our attention for a while to the other pillars of the system that is driving us all over a cliff: namely the corporate media – and the world of finance and business.

There was a small taste of the latter during our October Rebellion. We spent a day undertaking symbolic disruption in the City of London, targeting firms such as BlackRock, whose actions have been especially heinous in facilitating destruction of the Amazon. And we spent a day disrupting some flights at City Airport. Why City Airport? Used disproportionately by City-slickers, with its short runway making it suitable only for smaller (and therefore even more carbon-profligate) planes, City Airport is an obvious symbol of a dysfunctional system.

Does this mean that we are rabid Lefties with an axe to grind against capitalism? Not at all. You don't have to have any beef with capitalism in particular to understand that a world that allows unconfined growth in private jets is a world heading straight down the toilet. There just isn't the emissions space left for carbon profligacy. That isn't a 'left-wing' view any more than it was 'left-wing' in World War II to institute food rationing. Such egalitarian measures are simply what you do – simply commonsense – in the face of an emergency. We are calling for emergency-thinking to be put into action. This requires a reining in of extreme inequality, and it requires a reining in of 'the markets', not as a result of any ideology, but simply as a pre-requisite for common survival.

Perhaps you're thinking: but my company is a B-Corp, we're doing what we can, given market constraints; what more do these 'extremists' want? Sorry, but being a B-Corp really doesn't begin to cut it. If there is to be a future, please start thinking about far more radical changes that *you* need to make. Starting with this: join the other business leaders who have been brave enough to acknowledge that XR is engaged in noble and necessary work; give your employees clear permission to take time off to join Fridays for Future or XR events; declare that you will support employees who whistle-blow for common good or who get arrested trying to stand up for a future; give some real money to XR. We welcome anyone who is willing to take responsibility for changing the script. We are not 'anti-capitalists', as such, we are not going to scream in your faces; if we come and occupy your offices, we'll probably bring cake.

And above all, lobby government hard to regulate those companies who are racing to the bottom, who are fuelling destruction and undercutting those companies (like yours, perhaps) which are trying to do the right thing. Insist that the government create a level playing field; insist that all the economic activity of this country (and beyond) be directed towards achieving stringent, rapid reductions in climate-deadly greenhouse-gas emissions and in habitat destruction. This should start with an immediate commitment to make company accounts incorporate material climate risks *in the numbers*. Mark Carney, Governor of the Bank of England, is rightly proud of having got these risks into at least the narrative portion of company accounts, but real behaviour change, real market revaluation, will not be achieved until the numbers change accordingly. Otherwise these narratives are little more than 'corporate social responsibility' (pass the sick-bag), and the biggest market failure in history just carries on getting bigger.

Finally, let's be clear that this isn't about a few bad apples. This is about *system*-change. So sure, some of us are more equal than others in terms of our responsibility for the direness of this crisis. Some of us are going to have to change more than others. Expect to have a worse time of it this coming year if you are BlackRock than if you are John Lewis. But we are also truly all in this together; we all have to change. So don't expect that even being a leader in your field will immunise you against the demand for further major change. Remember that XR started making its name by occupying the offices of Greenpeace – to make the point that even the previous leaders in our own field had not gone nearly far enough in pushing for change.

So, if, at some time this year, you are upset because your boardroom has just been invaded, remember: we're your best friends. We're your best friends because we are the ones delivering ever-increasing pressure to help you demand of your company Chair – and of the government – what's necessary. And we're your best friends because it really *is* necessary. This system is coming to an end. It will either end as a horror story, in an uncontrolled collapse event, or it will end intelligently, deliberately, making way for something better.

If you care about your kids, if you care about the next generation, or even if you only care about being able to grow old, please join us in helping the outcome be the latter.

CHAPTER 23

THE BILLIONAIRE REBELLION

In January 2020, I went to Davos with the 'XR Catalysers' group to seek to catalyse change at the World Economic Forum by means of 'internal disruption'. That is: we didn't do nonviolent direct action; but we told the truth in uncomfortable ways. It was by turns an educational, inspiring, useful, and nauseating experience. On returning, I co-wrote with Jem Bendell, a veteran of Davos in past years, this piece for The Ecologist. *(This is our preferred version of the piece;* The Ecologist *published a shortened version which left out some key points.) We focused on our disagreements with Micah White, a sometime advisor to XR, over how real change is likeliest to happen.*

One man's promise to put less than 10% of his wealth towards climate action generated a lot of media attention last week. Jeff Bezos is the world's richest man, so it sounded like it might make a difference. Could this be the sign of things to come, as more of the world's billionaires rebel against a narrow pursuit of profit, power, and fame to try to save humanity from environmental disaster? Or at least try to stop driving us ever faster towards that disaster? If so, could seasoned activists like us find useful ways to engage them to generate a quantum leap in impact? After all, the situation is bleak, and everything should be considered. It is this seductive idea that led your authors to spend some time mingling with world elites, at Davos and elsewhere, hearing their views and sharing our own.

Our conclusion?

There is no shortcut to global social change. To transform systems, climate activists must focus on building power and support amongst diverse communities, and only welcome billionaire support if it is specifically for such empowerment. That is because sustained climate action will require a fresh settlement on the fairer distribution of resources, as we face a very challenging future. Such redistributions of power and resources have never been achieved merely by enlightened elites handing over what they are accustomed to.

A stark example of how billionaires and their global gatherings are not often inclined to such 'radical' agendas is the way the World Economic Forum (WEF) has engaged with the climate emergency. It is welcome that the WEF have now sounded the alarm on just how bad human-caused climate change has become for humanity. But that does not mean they already have significant suggestions for societal change.

This year at Davos they launched a manifesto with the title, 'The Universal Purpose of a Company in the Fourth Industrial Revolution'.[66] It contains warm words about some corporate efforts towards a future for humanity. But we can't see anything in it about firms (e.g., Facebook) not undermining democracy, or action on *the* causes of our time: social and ecological justice. The Davos Manifesto ignores the diminution of democracy (at a time when we direly need it to be deepened) that corporates have midwifed. Nor can we see anything in the document about the necessary role of the state in much-needed market interventions (e.g., the 'Green New Deal', aka the green industrial revolution) to address climate breakdown. Thus, this 'Davos Manifesto' also ignores what is agreed by most observers to be an absolutely key part of any genuine solution to the vast problem defining our time.

The rising tide of climate chaos that we see now almost constantly (most recently, in Australia's unprecedented bushfires, which killed a billion animals and poisoned the lungs of millions of people) proves the corporate model of running the world for profit has failed. The best

66 https://www.weforum.org/agenda/2019/12/davos-manifesto-2020-the-universal-purpose-of-a-company-in-the-fourth-industrial-revolution/

'manifesto' from Davos would be to take their members' money out of politics-as-usual and corporate media, and let ordinary people decide how to respond to a devastating global crisis that current elites have presided over.

A notable feature of the event this year was the call by Micah White, a co-founder of the Occupy protest movement, for the world's movements for justice and democracy to seek an alliance with the world's elites in order to address the emergency. One of us, Read, was seated beside Micah when he made his pitch for this 'alliance of opposites'[67] – and was one of those who argued passionately in that meeting that it would be simply impossible for ecological breakdown to be stopped in a world which allowed economic inequality to persist at its current level. For if a few of us live like there's no tomorrow, then there just won't be enough for most of us to have a tomorrow.

Micah's strategy and vision is fascinating, but on this crucial point and the non-negotiable need for a profound deepening of democracy (i.e., Extinction Rebellion's Demand 3), there seems to be a gap between his ideas and XR's. For instance, the Davos set believe that we should mobilise our movements to plant a trillion trees. Whereas we join our fellow climate activists in wanting to change the economic system that trashes forests and doesn't incentivise their planting. Some elites think we can leave the existing distribution of wealth and power intact *and* save the world. Most people we know in XR see that as wishful thinking.

What is a pragmatic approach in the face of an unprecedented emergency? Our view is there is no possibility of the crisis being tackled while people like those at Davos retain their wealth. We don't just need 10% of their wealth spent or lent for climate action. We must change economic systems so that there is a fairer distribution of limited resources as we collectively strive for net zero emissions and support each other as climate change disrupts our agriculture, water, cities, and health.

67 You can read Micah's account of that semi-confidential meeting here: https://www.independent.co.uk/voices/davos-2020-occupy-wall-street-trump-greta-thunberg-climate-extinction-rebellion-a9301131.html

Just a quick look at current climate impacts shows that redistribution is essential right now. For instance, in Kenya, climate impacts have led to fresh vegetables doubling in price in the past months. Similar situations are being experienced around the world. Faced with such a situation, there are even calls for tax cuts on oil and subsidies, to help bring down the price of food. That is the self-reinforcing disaster that will spiral out of control if emergency social justice is not at the centre of the climate agenda.

Climate activists can respond to this social and economic dimension of the struggle by forging more alliances across sectors and classes, including with trade unions, networks of school children, faith institutions, and others. Such alliances will be necessary not only to challenge current power but also to maintain new formations of power to deliver the massive changes required. One example is the need to engage trade unions, so that more of them decide to make climate safety as important to their bargaining and strike action as pay and conditions. When that happens, governments will begin to see the massive economic disruption that will occur if they don't make a climate justice agenda their foundational policy platform. The result of such alliances and policy changes will present a direct challenge to the wealth, privilege, and power of elites.

We say this without any rancor towards the wealthy. We say it simply because it's fact. It's *because* we care about everyone, including the super-rich and their kids, that we say to the Davos elite: it's time to give up your vast wealth and privilege. Let your money go: allow that money to be devoted, no strings attached, to the effort to change the world so that we all have a chance to survive on a more level living-field.

We think it is improbable... Unfortunately, it is not likely that corporations and the rich in general are going to swing behind radical action on the eco-emergency. But it seems to us realistic to aim to reach a few of them – those who have truly understood the science and the complicit system – to become true allies of the now-necessary radicalism. Imagine if we found an eco-equivalent of the Koch brothers. Imagine if, better still, we managed to get, say, 3.5% of the super-rich onside, on the side of reality, in a 'Billionaire Rebellion'. Such eventualities would be completely game-changing.

That's a reason why, if one can keep one's head and heart, it's perhaps worth going to Davos to invite a resonant reaction. If a handful of billionaires recognise that the system they enable is wrong and cannot continue, and begin to support the global grassroots climate movement, then we could see more rapid change. Together, we might yet respond to our terrifying predicament with as much love, determination, and courage as we have ever found.

CHAPTER 24

THE REPORT THEY DIDN'T WANT YOU TO READ

XR UK's newspaper The Hourglass *ran for eight issues. My regular column was called 'Another green Read' and included this piece below,[68] concerning a report produced for clients (not for public consumption) by J.P. Morgan. By a peculiar process that I am not at liberty to disclose, I came to have a copy of this report. I was astounded by its prognostications – by the extraordinarily frank assessment that it contained of how dire our common predicament is – and in February 2020 I began to leak them to the world. J.P. Morgan attempted to suppress the story, but the cat was out of the bag; journalists cottoned on, and for a while it became a big story. (My favourite story written about the report was Kate Aronoff's for* The New Republic, *which bore the magnificent headline, 'The planet is screwed, says bank that screwed the planet.'[69])*

The really good news is that, less than a fortnight after this report was leaked, J.P. Morgan announced a gigantic raft of reforms, including the end of financial support for drilling in the Arctic National Wildlife Refuge.

'It is clear that the earth is on an unsustainable trajectory. Something will have to change at some point *if the human race is going to survive.*'

68 https://rebellion.earth/wp/wp-content/uploads/2020/02/JPM_Risky_business__the_climate_and_the_macroeconomy_2020-01-14_3230707.pdf.pdf

69 https://newrepublic.com/article/156657/planet-screwed-says-bank-screwed-planet

Who do you think wrote those two terrifying sentences? Who is responsible for this latest exercise in radical truth-telling? The University of East Anglia? Caroline Lucas? XR?

The correct answer is: the world's largest funder of fossil fuels, J.P. Morgan, in a detailed report written by two of their leading economists – a report that they tried to pooh-pooh after I tweeted about its contents. Luckily the story couldn't be held back and has now gone worldwide. Those pesky eco-extremists at J.P. Morgan have blown the lid right off the debate and made starkly clear how close to the precipice we are. Perhaps some of those who don't like listening to you or me or Greta or even David Attenborough will be woken up by the deeply worrying words of a profit-hungry trans-national investment bank. For this is most definitely a bank that does not want to be letting such truths slip.

The thing that strikes me most strongly about the affair is J.P. Morgan's oh-so-telling response to their own report surfacing. Their denial that the report is a genuine J.P. Morgan report (when it contained no such disclaimer, and had their branding all over it), is the opposite of what they should have been saying: which would be to *own* this excellent report.

The reason they aren't keen to come clean isn't hard to figure out. J.P. Morgan are a huge and largely unrepentant funder of fossil fuels. If they were to accept what their own report is saying, that would require them to revolutionise their business model. It would require them to tell the truth – and act accordingly.

They want to hold out against doing this a little while longer. But the genie is out of the bottle. How long will institutions like J.P. Morgan be able to carry on complicitly with climate crimes now that their own economists are being clear as day about the consequences of doing so? It is quite obviously in the public interest that reports like this be made public. In fact, any information about our growing collective vulnerability to the biodiversity and climate crises ought to be disclosed to us.

So I want to say to anyone reading this who is privy to such information: please bring it out into the public. If you need help or protection to blow the whistle, then there is a platform available that supplies exactly that: www.truthteller.life.

For we need the full truth to be told. We need to know what army generals and military intelligence officers are scoping out in terms of contingency plans for the coming disasters and crises. We need to know about the vulnerabilities of our ludicrously fragile just-in-time food system. We need to know what the insurance companies know about the rising risks that they (read: we) are facing.

And so I hope that, in 2020, some of XR's actions will increasingly focus upon teasing out these vulnerabilities. Insurers that are not coming clean about what they know of the uncertainties and risks from which they seek to profit; supermarket supply chains; the governments overseeing all this: these are worthy targets for XR nonviolent direct actions that aim to highlight our vulnerabilities and those who are responsible for them. If such actions elicit whistle-blowers, so much the better.

Let's seek to dig out more reports and files and truths that they don't want us to read. That's how awakening proceeds: one revelation at a time.

CHAPTER 25

THE URGENT NEED TO TAKE CARE: FROM CORONA TO CLIMATE

This 'Another green Read' column for The Hourglass *was published early in the course of the UK lockdown to deal with Covid-19.*

I'm writing this, and you may be reading it, under 'lockdown'. Theoretically, this lockdown shows that our government is perfectly capable of acting in an emergency, when the mood takes them, when the emergency is in their faces enough. But does it?

Consider the following two points:

1. The reason they have implemented the 'lockdown' is that they were subject to relentless pressure for weeks from experts and commentators (including myself and my colleague Nassim Taleb of *Black Swan* fame[70]) who destroyed the absurd epidemiological models they were using that would have infected most of us and demolished our healthcare system, who challenged their complacency, posed alternatives, issued warnings, noted forcefully what was happening elsewhere, and so forth. In other words: together, *we*, citizens and experts alike, forced the course-change on the government.[71]

70 See in particular Taleb, et al., https://necsi.edu/systemic-risk-of-pandemic-via-novel-pathogens-coronavirus-a-note .

71 My claim that it is external public pressure which brought about the lockdown decision has since the time of publication been dramatically confirmed by admissions from within government: https://bylinetimes.com/2020/05/01/governed-by-opinion-governments-and-the-public-mood-in-a-crisis/

2. While they were prevaricating, we the people were already, in many, many cases, ahead of them – and so managed to keep ahead of the virus. Lives were saved by this; lives were saved by citizens choosing to act precautionarily: before the government issued any mandates at all, we voluntarily cancelled many events, shut down many institutions, started practising physical distancing, etc., etc. In other words: *we* led the course-change. We moved before they did.

These two points mean that if – and it remains a very big 'if' – the UK now manages to avoid descending fully into the hell that has overwhelmed north Italy in the last fortnight, if we manage to avoid most of our health service being completely overwhelmed – with the huge further spike in deaths and suffering that such overwhelm brings – then it will be because the citizens led the government. Not the other way around.

For even in the case of Covid-19, with the emergency breathing for just a matter of weeks down the government's neck, they were unwilling to act adequately to protect us. This bodes ill for their capacity to do so in relation to the far longer climate and ecological emergency.

It's just us, the people. We did this; we brought the UK to the point of having a shot at suppressing this pandemic. This is vital context for the period of community mutual aid that we must now enter into. A period in which there will be much need for quiet heroism, to save lives and reduce isolation.

We are *#alonetogether* in this struggle. Sitting in our homes, working on getting food to neighbours who need it, exercising at respectable physical distances from each other: we are #alonetogether. We express our mutual care at this time by phoning and not hugging. We the people have led; the government is dragged along unwillingly behind what we do, what we want, and what we see the need for and call for – and what we are co-creating.

This is vital context for the task of continuing to insist that the government do more – and that it doesn't move in the wrong direction. (E.g., it would be moving in the wrong direction now to pour resources into fossil fuel companies, airlines, or into the ecosystem-destroying proposed high-speed train route HS2, when those resources should be poured into making PPE and ventilators and into the pockets of those who must continue to work and may become sick, and so forth.)

Each in our own homes, and behind our masks, we are powerful, and we are together. We have led, and we need to lead more. The lives of our elders and medically vulnerable, perhaps our own lives, and certainly our self-respect, all depend on it.

The need to take care, to look before we leap, is something that the government failed to exercise when it plumped for 'taking it on the chin', and so lost precious weeks with which to hold back the incoming Covid-19 public health disaster. When we don't know something, we ought to protect ourselves against what we don't know. In the case of this emergency, this virus, it's unprecedented; we don't know what effects it has. We don't know, for example, whether it might have neurological effects, whether it might leave permanent lung damage. We don't know a great deal about how it's transmitted yet. And when we don't know that kind of stuff, we need to err on the side of safety. That's what it is to be precautious – and that is exactly what the UK Government has not been doing. In failing to impose travel restrictions and quarantines, in failing to lock down, in failing to mass-test, in failing to choose life over super-short-term economic business as usual, they haven't been keeping us safe.

The moral of the story? We are probably going to need to rely on ourselves in any emergency situation we face, now and in months and years to come.

Unless, perhaps, we can persuade the government to learn from its deadly mistakes.

CHAPTER 26

THESES ON THE CORONAVIRUS CRISIS[72]

This is my most sustained effort to tease out the meaning of the Covid-19 crisis. Unpublished in this form before, it accentuates the positive: this is the kind of spirit in which we might harvest the benefit from this terrible time of trial.

As the coronavirus crisis escalated, XR UK decided to suspend its planned May-June Rebellion. I wrote a document for XR's internal use, 'Some Strategic Scenario-scoping of the Coronavirus–XR Nexus',[73] which has fed into this

72 Huge thanks to Victor Anderson, Ed Gillespie, Jem Bendell, and to Extinction Rebellion colleagues including Gail Bradbrook, Skeena Rathor, Marc Lopatin, Joel S-H, Sarah Lunnon, and the XR Writers' Group. I alone take responsibility for any infelicities or fails in this piece, but its production was in a very real sense a team effort. Like anything worthwhile, the truth is that this piece could just as wisely be regarded as co-authored by a community as written by me.

73 That document can be viewed here: https://drive.google.com/file/d/1xeXfbu8oC9al_LH1NanaY8UZ2pJbXOxe/view?usp=sharing It was discovered by an enemy of XR, the 'journalist' David Rose, who wrote an attempted hatchet-job on XR and me for *The Spectator* about it: https://www.spectator.co.uk/article/revealed-extinction-rebellion-s-plans-to-exploit-the-Covid-crisis Rose's piece gravely misrepresented the contents of the scoping doc. In particular, he tried to make my discussion of how this crisis shouldn't be wasted into something somehow scandalous, when actually it is an entirely standard conception within politics and government – it was famously spoken by Rahm Emanuel, former White House Chief of Staff, on the last comparable occasion (the 2008 financial crisis, although the germ of the quotation is often attributed to Winston Churchill in the context of the final stages of World War II). What I meant by using the phrase is that it's quite obvious that it would be a gross collective dereliction of duty if we were not to learn from this coronavirus. The crisis it has imposed upon us should be used to ensure that we make ourselves less vulnerable to future crises: whether future pandemics, or the climate crisis, or whatever. It would be stupid – criminal – to let this crisis go to waste, by not preparing, through it, for future crises. That's just common sense. The story was then bounced elsewhere around the alt-right echo-chamber, including in this libellous piece by the execrable James Delingpole, a long-time critic of mine (a minor badge of honour), in Breitbart: https://www.breitbart.com/europe/2020/04/22/greens-celebrate-coronavirus-lockdown-as-blueprint-for-new-world-order/

piece. That document sought to explore how a direct-action movement focused on existential risks to us and our planetary home could act in a time of increasing restrictions on public gathering and movement. This piece looks wider, to how this moment might yet be the making of us. Us: humanity, having found some unity in this crisis that we hadn't yet managed to access in the 'normal' times of the longer climate and ecological emergency.

1. As I write these words, in May 2020, people across much of the world are experiencing a sense of prolonged personal and social vulnerability. Of lived emergency. For many of us, especially in the Global North, this is almost completely new. While it may build on experiences we've had, for instance, of worsening incidents of flooding (i.e. of climate disasters), it goes considerably beyond them. For it touches or threatens to touch us all.

2. And thus there is a radically new experience for most of us across the world: a sense of equality, of a radically *shared vulnerability*. (The last thing that was anything like this was the nuclear threat during the 1960s and 1980s. Covid-19, however, is breaking through to a level of globally shared awareness that even that never achieved. The climate crisis has certainly never yet achieved the same level of emotional and intellectual 'buy-in'; see below.) That equality isn't absolute, of course: The poor are suffering in greater numbers than the rest of us due to already-compromised health and less access to care. Furthermore, we also have ugly spectres like the super-rich diving off to bunkers, or buying for themselves coronavirus testing kits desperately needed by healthcare providers.[74] But those kinds of behaviours are widely seen to be unacceptable, just like the prime minister's Chief of Staff brazenly breaking the very restrictions he had imposed on the rest of us.[75] The attempt to buy one's way out of the vulnerability we now face together is an evasion. Even the mightiest are vulnerable: thus

74 https://www.mirror.co.uk/news/uk-news/coronavirus-harley-street-clinic-sold-21711100

75 https://www.mirror.co.uk/news/politics/boris-johnsons-defence-dominic-cummings-22079171

the British prime minister, his Chief of Staff, and even the heir to the throne all contracted the virus, the former ending up in intensive care. We are collectively humbled by a microscopic virus. This sense of vulnerability is what we in XR have been trying to evoke. Our thinking has been: if only we could get people at large to feel the vulnerability that they really are subject to, and that we feel. In this striking sense, to put the matter somewhat crudely, Covid-19 has done XR's job better than XR itself could have hoped. This horror that we are now living alone-together is the mother of all wake-up calls. Our collective vulnerability is literally coming home to us.

3. There has been a vast failure of governance in countries like the US and UK in this outbreak. A failure to observe the precautionary principle,[76] a failure to value citizens' lives and health above crude ultra-short-term economic imperatives, a failure to protect. This is a depressing fact, because the Covid-19 crisis should have been much easier to address than the climate crisis: because its imperatives are far shorter in timescale, its damage far easier to see and to attribute. But even this failure has an upside. We are realising that our governments are not going to save us, but that we can move ahead of them to save ourselves, to save each other. And these governments may emerge from this crisis brittle and vulnerable. There is thus a greater possibility than there was in 2019 of moving decisively beyond them. For we can now say plain to such governments: You failed to follow the common sense that is precautionary principle when it came to the coronavirus…why on Earth should we trust you when it comes to the climate crisis? Those

76 I made the proposals embodied here https://medium.com/@rupertjread/what-would-a-precautionary-approach-to-the-coronavirus-look-like-155626f7c2bd available to the UK Government in February 2020. Like so many other sound precautionary proposals at the time, these were ignored. Other island nations, such as Taiwan and New Zealand, by contrast protected themselves adequately against the emerging pandemic. See also https://bylinetimes.com/2020/04/23/the-coronavirus-crisis-documents-reveal-government-and-nervtag-breached-own-scientific-risk-assessment-guidance/ & https://medium.com/@ian_js/a-national-scandal-a-timeline-of-the-uk-governments-response-to-the-coronavirus-crisis-b608682cdbe

governments which let their peoples down by being unprecautious even regarding the immediate threat posed by Covid-19 – by allowing hyper-mobility to continue even when it was clear that we were heading towards a pandemic, by prioritising economic growth over life itself, by failing to provide sick pay for those in the gig economy etc., who therefore were incentivised to work sick and keep spreading the virus, and so on – most certainly cannot be trusted to deal with the more diffuse yet ultimately far more cascading, terminal and total threat of the ecological emergency. Covid-19 has been a dry run for the emergency: governments such as ours did not succeed in that dry run. We ran the experiment once, which is all you ever get to do in the real world; we (rather: our government) fucked up royally; so now it is time to trust citizens and experts instead.

4. The radical newness of the moment we are inhabiting may well change our collective sense of history itself. The outbreak is unprecedented – there has never been a pandemic with the capacity for massive mortality in our globalised time; and never before in our lifetimes has our sense of 'progress' been challenged in this way. Time may well be divided into a new BC and AC: Before Corona and After Corona. Any attempt to blindly return to normal after this will be haunted by this question: What is 'normal' now?

5. The shared vulnerability we are experiencing could be reacted against by a retreat into separate silos, a conceptual echo of our current physical distancing. Or it could propel us into an emerging global consciousness that brings us together in our vulnerability and that manifests a more beautiful world that just possibly is starting to become actual. The call that the Covid-19 crisis makes upon us is surely to the latter: because it shows us that my health is your health; that there is no health without public health. The challenge it calls us to, surely, is to make this emphatically and unequivocally a trans-national, trans-political issue of empathy. And to carry that sense of empathy forward

into the longer emergencies that will be here even after a vaccine for Covid-19 comes along (assuming it does).

6. This moment, this tragic and horrific event, this shared emergency and this arising consciousness, is changing what is understood to be possible. Further, it is already changing what is. In bad ways, and good. And the way it works on people is going to be complex – and sometimes surprising.

7. Many people – those who survive and who are not crushed by the grief that is engulfing some of us – will likely recall this time positively. Because suddenly their lives have meaning; because they know what/who they love, perhaps for the first time. Many survivors of World War II later recalled it as one of the best times of their lives. Counter-intuitive, but true. There's an 'apocalypse' moment here in the deepest etymological sense of the word. 'Apocalypse' means 'drawing back the veil'. What we are going through reveals the 'reveal' in revelation. Where we see and feel what really matters; perhaps for the first time.

8. Thus there really is, I think, the possibility of a positive through-line here, one which centres on us learning, in this emergency, what really matters. Love matters more than 'growth'. Your father matters more than another point on (or off) GDP. Human life matters more than 'the economy'. If that truth lands, it will have transformative power.

9. Disasters can bring out the best in us.[77] Disasters call upon us to be our best. They often lead to us suspending normal rules of decorum or of self-interestedness, and to our exhibiting altruism we didn't even know we were capable of. We are seeing #everydayheroism aplenty. We are building paradises in the midst of hells. Disasters are disasters; but they invariably contain the possibility of a silver lining, if we are ready to

77 See https://www.resilience.org/stories/2019-10-29/disaster-localization-a-constructive-response-to-climate-chaos/

make it. This one is a hard disaster to make something good from, for sure – especially because we can't gather together in the way people did during, say, the Blitz. But we're finding ways of making it happen, all the same; not just online, but through mutual aid, and more – read on...

10. Covid-19 is showing us that we are good! Sure, there has been stockpiling verging on hoarding, but let's not be too judgemental of that. In this unprecedented situation it is natural to feel the fear and to feel concern about your own larder. It shows in fact a healthy awareness that this is for real – which, if it carries forward to awareness of our vulnerability to interruptions by climate disasters of a fragile just-in-time food system, will be healthier still. Especially if we then get serious about building redundancy and buffering into what is an overly 'optimised' system. Furthermore, it shows a realistic appraisal of the lack of guarantees from a government that would have been able to reassure people if it had implemented, say, a food-rationing scheme. In any case, Covid-19 is showing us the very best of ourselves, above all in the selfless professionalism of health care workers and in the marvellous outpourings of love for those workers. When I went out to my front doorstep on 26 March at 8pm to #clapforourcarers, I was afraid there would be a poor turnout. Maybe everyone else had the same fear. But we needn't have done. Most of my street was there, on our front doorsteps and at our windows. It was so moving, such a tremendous experience of community. This gives real hope. For if, in this hard disaster, where physical distancing can make us feel a little paranoid even towards our neighbours, we are still succeeding in building community more than ever before, then how much easier that community building could be once the pandemic is over and we can literally stand shoulder to shoulder to tackle the enduring challenges of our age. My own experience in the pandemic has been one of experiencing physical distancing as itself a manifestation of community and of care. When I've been out for my daily exercise, I've found a newfound fellowship with others in keeping two metres away from

them. Paradoxically but truly, there's been more good will, more togetherness, in our keeping apart than I've felt for many, many years in my community. My sense as we've kept carefully apart from one another is that precisely in that care we are in this together. And this is something that my heart yearns for.

11. This radical experience of community will not just fade away. Sure, some with vested interests in returning us to the old growthist treadmill will try to make us forget it; sure, some of us will want to forget it, because of the pain of this time and also because of the hope that was born in this time. For hope is painful; hope renders you vulnerable (see Thesis 22). It will be tempting to forget. But this is history we are living through. This is the meaning of the human story morphing, right now in real time. This is a moment that will live in our memories, like it or not. This radical experience of community is with us now, part of what we cannot help knowing to be possible, real, literally vital. For Covid-19 forces us, wonderfully, to think like a community. In the best possible sense, the coronavirus actually does force us to think like a 'herd'. In the early stages of a pandemic, it is tempting to think like an individual: 'the chances of me suffering from this, at least at this point, are very remote. The chances of me carrying this and infecting others, at least at this point, are likewise very remote. So I'll go on acting as I normally do for now.' But if everyone thinks like that then it guarantees the pandemic will spread like wildfire – and that soon it will no longer be true that the chances are remote. What is needed is for everyone to think towards the community level from the beginning. And that is how we have been learning to think. For in countries where we haven't, we've noticed the cost.

12. This pandemic could claim millions of lives, and leave millions with damaged health.[78] We do not yet know the possible extent of the global tragedy that will unfold in the coming months (and

78 https://www.telegraph.co.uk/global-health/science-and-disease/coronavirus-could-cause-secondary-illnesses-including-chronic/

years). The lockdown of entire countries gives us all, however, an opportunity to pause and to start to reassess the sort of society we have created: the global society that co-created the Covid-19 pandemic, through its destruction of habitats and its wildlife markets (see Thesis 24), and through its hyper-mobility (see Thesis 13). We should take this opportunity to reflect on the blindness and even hellishness that have been part of and consequences of our 'normal' lives and systems. And to reflect on the fabulous things we have been taking for granted, before this pandemic: like being able to hug friends, and being able to take holidays in nature.

13. And we should reflect on the extreme fragility of our economically globalised world. The fact that a virus can spread so fast and shut down large sections of the global economy in a matter of months should deeply worry us. (This is especially relevant in the context of catastrophic climate change, which threatens to do much the same but on an unimaginably magnified scale.) If we want to get serious about minimising harms like this, then we should scale back economic globalisation and reduce the extent to which countries require international trade and travel. Countries and localities should have food sovereignty, be less dependent on others, be more resilient, be less physically interconnected. Producing more stuff on a local or regional level will inoculate us against the types of supply-line disruptions that we can fully expect catastrophic climate change to bring – disruptions that we are about to experience some of, due to Covid-19; a further development of our sense of shared vulnerability. Reducing international travel will also reduce climate-deadly emissions, of course, just as the virus already has. What this is all part of is a revival of the local – something that many across the political spectrum are hungry for.

14. However, a less economically globalised world does not mean a more Balkanised or nationalistic world. Moving to stop international air travel that is carrying the virus, moving to implement

quarantines and lockdowns: these are not autocratic actions, they are intelligent actions for public health. As noted in Thesis 5 this corona crisis is all about interconnectedness, interdependence, indivisibility – but achieving these things need not equate to going places; as we are experiencing, under lockdown. Likewise, let's preserve and enhance our sense of mutual union while relocalising our systems. Let's stay in touch across the world online while radically reducing the movement of goods and of people. We need more political co-operation – international co-operation – on medical responses to the pandemic (as well as on other things such as the biodiversity and climate emergencies); we are realising, slightly late in the day, that organisations such as the World Health Organisation are (literally) vital. But that needn't involve old-fashioned summitry with leaders flying in and making pledges on climate that their own behaviour starkly contradicts. Our arising global consciousnesses can enable us to understand how we help each other by all making our places and spaces less physically interconnected. We practise care for each other by reducing the extent to which we engage in pointless international 'food-swaps', and by increasing the extent to which we look to strengthen bioregional economies and polities.

15. Lockdown isn't easy, it isn't what any of us would have chosen. But millions of people and workplaces are beginning to discover that the amount of work that can be done at home is far greater than is often assumed. In-person meetings can often be replaced by well-worded emails or videoconferencing. These are preferable from an ecological perspective, more efficient timewise, and much more family friendly – allowing people with children to work from home. The reduction in air pollution that such measures lead to would also vastly reduce respiratory disease, making people less vulnerable to viruses like this in future. There are humungous costs to a world whose wheels are oiled by normalised long stressful daily commutes: tens of millions of wasted hours every day. Once the lockdowns end,

we should seek to preserve the positive lessons we've learnt about how we can live differently, not just try to flip back to 'business as usual'. My hope is that this global tragedy can lead to seeds of positive change. In ways that are already hinted at in theses 2, 5, 13 and 14, I hope this will prove to be a sustained wakeup call to how poorly prepared the world is for the climate crisis that we are hurtling towards. And I hope that right now we stay safe and support our communities. Because without that sort of solidarity, emergencies will always be worse than they need to be. I believe that these hopes could truly be realised. This is a moment of testing. We are being tested by nature (see Thesis 24) and (inadvertently) by ourselves. Will we become smaller or rise with solidarity to the challenge?

16. The answer to that question depends on how the struggles over the meaning of this emergency unfold. Consider some oppositions that Covid-19 makes stark: people vs profit. Bailing us, the people, out vs bailing out corporations and carbon. Love vs desperate attempts to prolong business as usual. Love for the vulnerable and rage against the heartless machine vs thoughtlessness and an unwillingness to think beyond what we knew 'B.C.'. Now is the time to decide whether to be real, or to try to shut out the facts that are battering on our minds with an insistence that will not be denied. Now is the time to get serious in asking how and why we have allowed ourselves to become collectively so vulnerable; and how we can care for our most vulnerable right now and in the longer term.

17. But in a way the decision is made for us: business-as-usual is not possible. Radical change is coming. In fact, it is already here. A.C., the world will never be the same again. Never again will we be able to simply ignore the costs, the silent risks, of the project of economic globalisation. Never again will it be possible to pretend that our love for one another is something marginal.

18. Never again will it be possible to pretend that there isn't the money to make things better. Covid-19 has made wildly ambitious ideas, way beyond the pale of political normality – such as stopping the destruction of biodiversity and ending climate-deadly carbon emissions in rich countries by 2025 (the second demand of Extinction Rebellion) – suddenly seem attainable: because we have seen governments move heaven and earth, with unimaginably vast economic packages, in the space of just weeks or even days. Covid-19 has made the 'politically impossible' necessary. Why stop at 'mitigating' the worst effects of climate damage? Covid-19 has given us a wonderful new word: suppression. We ought to be suppressing climate-deadly carbon emissions as fast as we humanly can while effecting a just transition. So why not do it by 2025, to keep ourselves safe? And this directly implies changing everything, as we emerge from this corona emergency: after all, when your house has burnt down, why would you rebuild it as before? We need to rebuild together the house of our dreams, with love at its heart.

19. The deep learning that is possible hereabouts centres, I believe, not on the availability of vast economic 'stimulus' packages but on the following proposition: we should prioritise people, not 'the economy'. The economy is here to serve us: not vice versa. We should practise people-protection. Because people are experiencing their vulnerability, the vulnerability of those they love – it is possible that, in those countries where the response to the virus was botched, virtually everyone will eventually know someone who has died from this virus. And many of those who have died will have died because governments such as those in the US and UK prevaricated[79] before taking serious precautionary action on Covid-19 – apparently because they

79 See https://www.facebook.com/notes/rupert-read/open-letter-to-uk-government-from-the-editor-of-the-lancet-chris-packham-george-/10163114716620301/ for the kind of thing they should have done.

couldn't bear to shut down economic business as usual. We see now that what 'the economy' really comprises is *people* who are vulnerable.

20. Of course, 'the economy' provides most of people's food and livelihoods. But that can change. The economy as we know it dominates much of people's lives, but we could have a provisioning economy which is more limited, devoted to giving people what they need to exist. After all, we are learning more about what we actually need during this crisis; how much less we need than what the advertising- and marketing-fixated economy tells us we need. Or, drawing on Thesis 8, we could go further, as my friend Chris Keene suggested to me: Humankind existed for hundreds of thousands of years without an economy. All we need is water, food, shelter, clothing, medicine and energy … and connectivity to keep us all sane in lockdown. The government could organise the provision of all those things without using money. Similarly, food rationing may yet be coming to countries like the UK, incapable as we are of feeding ourselves, at some point during the time of Covid-19[80] as an alternative to hoarding and possible breakdowns of social order, if some people cease to have enough food to live. And that's an example of how there is something that truly trumps 'the economy': life and our societal determination to preserve it (or otherwise). Of course, economic risk and hardship through this crisis are massive; of course, many people are going to be worrying about that a great deal too, now and for a long time to come. Worried about their jobs, their finances, etc. Indeed, we may be about to be precipitated into a global Depression. But even a Depression, if it is managed right, doesn't literally kill people, or at least, not vastly many. Whereas the exponential potential of the coronavirus to take out the old and ill in thousands or perhaps tens of thousands daily, to overwhelm our healthcare systems,

80 Because some key food exporters are banning exports: https://www.feedstrategy.com/coronavirus/some-countries-ban-export-of-staple-crops-amid-covid-19/

and thus to lead to a generalised public health catastrophe (and potentially a social order breakdown), is of a different order.

In the end, what corona will surely teach us is that GDP isn't all it is cracked up to be. When we remake the world economy out of this crisis, it will be essential to remake it in a way that prioritises lives and wellbeing.[81] The pandemic has taught us whose work society really depends upon: the health workers, the food producers, the cleaners, etc. – often on low pay, much lower than those whose work is not at all essential, such as those who wheel and deal in anti-social financial services. Let's not just flip back desperately to living the same way as before: to the air pollution, the noise pollution, the frantic commutes, the resulting fragilities that this virus has exploited. Let's refuse the kind of mother of all spending sprees that our governments are already planning to frantically reboot economic growth – unless that spending is designed to actually make things better. Let's remake our world in a manner that no longer worships that abstraction, 'the economy', as if it is a deity. And crucially, let's be certain to do this in a way that doesn't throw us straight from the frying pan of corona into worse fires. If we try to 'stimulate' the economy regardless of other social or ecological costs, we will prove that actually we learnt nothing of significance from this existential crisis. After 2008, we didn't put people or planet at the centre of our reconstruction efforts and look where that got us. We can't afford to do that again. Surely, through lived experience, this time we have learned. We've learned the hard way that there are no degrees of separation.

21. Once there has been this shared sense of emergency, of deep vulnerability, there is no going back. If as a species we can learn from this, change some of our practices, embrace precaution and ethics, build on the mutual aid and love we have shown

81 See https://planb.earth/wp-content/uploads/2020/03/Press-release-26-March-2020-final.pdf for an XR-ish picture of this.

during this time of corona, we will be forever changed. Imagine not only being able to hug your friends and parents again, but then holding onto the knowledge of that preciousness. Make that real. The way to honour the memory of those who have died and are dying is not to lurch back to the system that contributed to killing them. It's to co-create something better.

22. In Thesis 2 I suggested there's an equality in our shared vulnerability. This is true if we truly care about our elders (and about the medically vulnerable). If those among us who are lucky enough to have low (no one has zero) susceptibility to Covid-19 are serious about loving our parents and grandparents, our uncles and aunts, our more vulnerable friends, etc., really take to heart their vulnerability, then we are all vulnerable. For love makes us vulnerable – and this is a good thing. To paraphrase C.S. Lewis: the only place outside heaven where one is immune to that kind of vulnerability is hell. It's connected to the fact that the capacity to grieve is part of what makes us human.[82] There is no love, without, sooner or later, grief. We've been reminded, in this corona crisis, about what's important: our care and love for one another and having a system that makes that possible and lasting. We can build the future around that.

23. This makes possible a gift from this virus to our common future. We are, as Dougald Hine has said, experiencing a planetary crisis of parental mortality.[83] Those of us who are not in the highest age-risk bracket vis-à-vis Covid-19 are having to face, all at the same time, the prospect of potentially losing our parents or grandparents. And that potential requires of course a loving precautionary protective response. What love in the time of corona means includes not needlessly hugging, so as to preserve life. The young love the old in this crisis; through

82 See my philosophicalish picture of how this is so: https://rupertread.net/writings/2019/what-grief-personal-and-philosophical-answer

83 See this special episode of Dougal Hine's Notes from Underground podcast: https://www.listennotes.com/podcasts/notes-from/special-edition-the-price-of-vOfUpmd31ut/

mutual aid, through physical distancing, through just caring and loving, and so forth; the old will afterward have the chance to repay the favour. Can the old learn to love the young in the longer emergency of the climate and ecological crisis? Corona is a test-case; looming climate breakdown will be even harder to pass. Climate is so much harder than corona to deal with, for beings such as ourselves with narrow time-horizons. And yet, as hinted in theses 8 and 9: there's also a sense in which it is easier. For corona may sometimes make us afraid of each other, and so potentially pushes us away from each other into illusions of separation (even as we know that we are tied together in a common fate by the possibility of contagion – and by the public health sanity and forcefulness needed in order to counter that contagion). Whereas with climate, at least it is clear that we have to come together in the mother of all mobilisations in order to have any chance whatever of rising to the challenge. What corona offers us, in short, is the prospect of intergenerational reconciliation. Nothing could be more badly needed, for us to have a future. Once we've saved the old, we must save the young.

24. The Covid-19 pandemic has moreover shown that we destroy the natural world at our peril. Destroying the forest habitats of bats and pangolins (and caging them) pushes the coronaviruses that live on them to seek new hosts instead – in this case, humans. Coronavirus is a warning from nature[84] as well as a test. Will we heed the warning? Will we pass the test? That depends on us; it depends on me, and you, and everyone else, and on what we learn while we spend this unique time #alonetogether.[85]

84 https://www.theguardian.com/environment/2020/mar/18/tip-of-the-iceberg-is-our-destruction-of-nature-responsible-for-covid-19-aoe

85 See https://rebellion.earth/wp/wp-content/uploads/2020/03/covid-19_regen_handbook_FINAL2.pdf?fbclid=IwAR2Rk2wshd3UjgxEVGH7Pq1pnsUtPv6c5WS18mc-K7pOsqkohz3VDJDR6BU

25. Dangerous anthropogenic climate change will produce more pandemics.[86] It may already have increased the risk and played a role in the coronavirus outbreak. In this specific regard, Covid-19 really is like a relatively mild dry run. We could get something far, far worse down the line. We have to rein in the exponential expansion of industrial civilisation, the excessive land use changes, destruction of rainforests, encroachment on wildlife, and beyond. If we don't, this horror-story is only a taste of things to come.

26. It is very hard for human beings to imagine things radically outside their experience. A 'normalcy bias' makes us very poor at being ready for 'black swan' events. Uncertainty, 'fat tails', and precaution are little understood.[87] Crude, over-simplified versions of 'evidence-based' analysis predominate. There has been no true global pandemic with high mortality within the lifetimes of virtually anyone now alive, i.e., since the Spanish flu. And since then we have, as humans, become more and more pleased with ourselves, increasingly confident that our technology, science, and understanding are such that we are allegedly near-invulnerable to threats from the mere natural world. Such hubris comes before a fall. Humanity has stumbled in response to the coronavirus, but not fallen. If we can learn from the stumbling, then we might even now manage to avoid the fall that the ecological emergency represents. Doing so, as set out in Thesis 21, would be *the best conceivable way of memorialising those who died*, most of them unnecessarily, in the coronavirus pandemic.

86 See https://www.resilience.org/stories/2019-07-24/global-pandemics/ , penned before this pandemic.

87 On why, see my article at https://iai.tv/articles/pandemic-precaution-and-moral-obligation-auid-1383

CHAPTER 27

NOW IS THE TIME – OR ELSE OUR HOPES START TO FADE

This piece, not previously published, is a counterpoint to the previous chapter. Here, rather than accentuating the positive possibilities the start of the 2020s presents us, I take a ruthlessly realistic look at the prospects for movements like XR – and by extension for humanity – in light of how the Covid-19 crisis is so far playing out. This represents a new frontier in facing up to reality (at least for me!) and suggests just how challenging the state of play is for XR (and for humanity) as it moves into its new phase – 'XR 2.0', as some of us are calling it. If my '26 Theses' were a kind of prayer for what ought to happen now, this is my grittier prescription for adapting to the situation we are likely to find ourselves in.

As this book comes towards its end, let me start by making a reflective remark about the psychological nature of activism typical among those of us awake to the threat to our common future. We cycle through periods of hopelessness on the one hand and desperate hope-against-hope on the other. We veer between not seeing how we can possibly make it through the long climate emergency, and insisting that we will. Between being tempted to give up, and throwing our all into a courageous no-holds-barred defence of Mother Earth and of our children.

If you've ever been tempted to give up – and I know you have! – then 2020 has already become a year to remember. Covid-19 has proven that one should never feel too sure of the state of the world or of what is or isn't possible.

It's been the worst of times and yet, in a way, the best of times. We've had the chance to feel emergency together – the very vulnerabilities, supply-chain issues, the very sense of disaster *and* of unity that we in XR sought to evoke through our rebellions, have been scaled up for us in a manner more powerful than we could have imagined. Whole fleets of planes have been grounded, air pollution has fallen dramatically, where nature was banished she has begun to return. Britain's climate-deadly emissions fell by a third in one month! Together, to save our most vulnerable, we decided to protectively contract the economy. (Don't let them get away with calling it a 'recession'. It's a deliberate, wise, protective contraction.)

Doesn't your heart leap to contemplate it?

If we had thought it was too late, we've been given one, almost certainly *last*, chance to get things *right*. These moments of reset don't come around every year, nor even every decade. (The previous such chance was in 2008.) This is it, now.

But at the very same time we need to be realistic. The forces blocking a profound, full post-corona reset are impossibly vast. Financial and political power are gearing up right now for a good (sic) old-fashioned growthist rebound. There is already plenty of carbon bailing-out going on,[88] a bonfire of environmental regulations in the USA, and strongmen around the world (for whom ecology is not exactly a top priority!) making opportunistic grabs for more power. The digital behemoths are immensely strengthened; meanwhile, public transport is challenged as it never has been before, by the virus.

This is likely our last best chance and we must seize it. But I hate to break it to you – though I think that deep down you already know full well – *it's not going to be the full salvation we might wish for.*

The cycle outlined at the start of this piece needs to be brought to an end. This is very hard to accept. But actually, our realest hope lies

88 https://www.theguardian.com/environment/2020/apr/17/polluter-bailouts-and-lobbying-during-covid-19-pandemic

in accepting it. Through brave actions and absolute determination, we must achieve a reset that works for people and planet. But we have to do so while being clear in our hearts that we just are not going to win outright. That the best that we can hope for is to soften the blow that humanity is going to receive in the 2020s. Environmental scientists have themselves now started saying this.[89]

Remember that we face a potential dilemma: if our advocacy for a more equal post-growth future seriously threatens the lifestyles of the rich and powerful (or of ordinary people), it will likely fuel the 'populist' counter-reaction. Is there any honest scenario that enables us to win deeply, swiftly, and completely enough that we can realistically picture resetting from this crisis in such a way that we achieve carbon-zero and biodiversity-loss-zero in countries like Australia and the UK by 2025? Across the whole world, by 2030?

The International Energy Agency forecasts an 8% drop in climate-deadly carbon emissions this year. Amazing. This is it, that rarest of opportunities: the collective chance to reset, that we so desperately needed. But we need to maintain – in fact, to exceed – that 8% momentum *every year* for the next decade, while somehow accommodating or dealing with continuing demands from powerful quarters for the full, enduring 're-opening' of the economy. Is that really going to happen?

So: let's not pretend that the reset will be all we desperately hope for. In fact, if we were to think that way we would be setting ourselves up for burnout – and that, of course, is what the cycle I referred to at the start of this piece is really about. We would be setting ourselves up for another round of devastating disappointment.

What does this imply? It implies that anyone imagining that the reset from this crisis can be a full salvation requires some gentle, loving deflation. Fantasies of imminent triumph are not the best we can do. If we

---

89 In fact, some of them are going considerably further than me – check out this multi-signed letter in *The Guardian*: https://www.theguardian.com/world/2020/may/10/after-coronavirus-focus-on-the-climate-emergency

recognise them within ourselves – and who doesn't? – then it's time to take a loving look at them and start to let them go.

By being honest in this way, we gain the clarity to see what this rarest of all windows of opportunity actually offers us. It is not some glorious utopian victory. It is survival. It is a gentle descent from our current civilisational hubris, and wise adaptation to the climate and ecological decline baked into the next generation.

XR needs millions of people all around the world who are inspired enough by the upside of human nature that has been so visible in this pandemic, angry enough at their failed governments, and determined enough to seize this last chance, to come together in clarity to reset – in a rather different way to what we have been imagining.

The only way we find these millions of recruits is if we signal clearly that they can't rely on XR to save the day *for* them. Instead, what's needed is that spirit of true mutuality that has been so splendidly visible these last months. Those who have been close to getting on board with us need a final push: that push, paradoxically, is XR being straight with them that we're not going to pull off some magical victory. Straight talk achieves resonance and congruence where 'Pollyanna'-style dreams cannot.

There's a space between total defeat (the eco-driven societal collapse that's coming if the reset from corona is not sufficiently wise) and the victory that most of us harbour desperate hopes of. It's *that* space that the corona crisis opens up – of potential flourishing and resilience, even as we adapt to a world whose ecology will be continuing, on balance, to worsen for a long time to come.

Talk of phase-shift to some new civilisation where everything is going to be different and great is bullshit. The simplistic heroic story of 'Twelve years (or twelve months) to save the world' is hopium; a dangerous drug. A story we tell ourselves so that we can picture ourselves as heroes or martyrs or messiahs, while crowding out the messier truth.

Our world isn't going to be saved. But we can still stop it from being committed to destruction.

But we're already on the brink of losing our last best chance to prevent a catastrophic collapse event or events which will make the awful coronavirus crisis look small by comparison. We *will* so collapse, unless we can dare to be brave enough to admit that everything is not going to be OK. When citizens, the elites, and our potential activists and allies see XR evincing such honesty, it will create the space for them, too, to face the emptiness inside. By being in this way brave ourselves, we offer them the courage to admit that they are terrified and in despair too. For they, too, want to stand up and be counted. But so long as they can outsource to us their imagined salvation, then the existential crisis that the existential threat facing human civilisation should manifest will not happen.

This is a prime, vital chance for truth-telling. For acknowledging that the best we can now hope for is a kind of marvellous muddling through; and even that is going to take extraordinary courage, extraordinary willingness to change, extraordinary willingness to 'sacrifice'. We are going to have to act like we have little left to lose, as is true.

If I've sounded slightly repetitive, it is deliberate. The message is an un-easy one; it needs repeated contemplation.

When I contemplate the truth that it's time to give up the hope I've cherished all my adult life for us to win, I feel grief. This rare, desperately urgent reset-moment is, in truth, for something much less than we'd hoped for, something decidedly imperfect. And as a society or species we'll *still* probably blow it; we'll probably miss it, waste it … and, if we waste it completely, then we'll most likely be 'committed' to collapsing. That is deeply saddening.

Of course, even if that is the way things go, *a luta continua* – and the struggle will probably morph into one between better and worse ways of collapsing. That would still be a struggle utterly worth taking part in. It's never too late to do the right thing. 'Deep Adaptation', readying oneself and one's society for potential collapse, is worthwhile. Indeed, I've argued already in this book that it is deeply necessary. It is a crucial part of the adaptation process we now need.

But what a shame if we lose the chance to pre-empt Deep Adaptation from becoming the only game in town.

Owning up to that possibility, I feel the same kind of nervousness and loneliness that I did three years ago, when I started saying that this civilisation is coming to an end. I was horribly worried then that I would be called out as a defeatist, even a traitor, and that indeed the real-world effect of my truth-telling would be to demoralise the very activists who most need bucking up. But it turned out my worries were largely misplaced. The calling-out I feared hardly ever happened; it turned out that, far from condemning myself to a lonely ghetto, I had found my gang. And it was massive and had a plan for changing the world.

So I'm seeking to trust now, as I have learned to over time. To trust that this new version of my message will also find its resonance. To trust in what my spirit knows: that we must remain honest in acknowledging that a 'green industrial revolution' would not save us, and that in any case such a transition is not going to happen everywhere, fast enough.

The coronavirus has taught us to feel our vulnerability. Amidst its horror, that is an incalculable gift. We squander it if we race off into fantasies of new invulnerability. The virus has shown us the brittleness of our systems. There really is a huge opportunity here: to strengthen them. To relocalise. To stop relying so much on long, polluting supply-chains. To adapt! This will be transformative, if we succeed in it.

And to end on a genuine note of new hope: It's not just the likely last chance for transformative adaptation, it is truly our *best* chance. How so? Because lockdown has been the mother of all pauses. This is the first time in decades, centuries even, that many parts of the world, including cities, have seen peace and quiet. An appreciation for nature and for calm like we haven't known before has suddenly been democratised. Nature likes it, and so do we. Covid-19 has kickstarted a rewilding, a re-opening of space for nature and for the restoration of biodiverse places, and the allying to that of appropriate agroecological and permacultural methods in food-growing. Here is an opportunity as great as it is rare for a serious agenda to create systemic change.

If we can play a part in making that happen, that will be a powerful way of loving life indeed.

Our descent from the world as we've hitherto made and known it can be brutal, rapid, uncontrolled. Or it can be soft and wise. If enough of us understand that to choose a path other than feasible adaptation is to choose the path of collapse, then we can realise the promise of our movement. We can rebel successfully against extinction, we can considerably mitigate the ecological crisis, we can prefigure a culture that will be regenerative rather than destructive. We can manifest the spirit our time demands, beautifully, lovingly – and efficaciously.

What more could you hope for?

POST-SCRIPT

THE REBELLION HYPOTHESIS: CRISIS, INACTION, AND THE QUESTION OF CIVIL DISOBEDIENCE

Samuel Alexander

I am grateful for this opportunity to provide a closing statement to this collection of essays. One approach would be to review the content in the preceding pages, highlighting the points of agreement and exploring any points of disagreement, doubt, or uncertainty. But given that this is a relatively short book, a summary doesn't seem necessary, nor does it seem right to present a sympathetic critique of this issue or that framing, especially since readers will still be digesting the ideas and perspectives which are rich in provocations. Instead, I will offer a few thoughts of my own, in ways that I hope serve as a fitting conclusion... to what may really just be beginning. As Rupert suggested in the preface, the Covid-19 disruption could well mark a coherent distinction between the first phase – Extinction Rebellion 1.0 – and the post-Covid era – Extinction Rebellion 2.0, where the world will be, and should be, especially sensitive to the social justice dimensions of any bold climate response.

Extinction Rebellion, as we have seen, has emerged as one of the most active and prominent faces of the environmental movement around the world. While a 'protest' on a particular issue may come and go, a 'rebellion' defines itself by the breadth of its opposition and the refusal to fade away, even in the face of slow progress or backlash from the state – or even in the face of a pandemic that prohibits mass mobilisations in the streets. Whether Extinction Rebellion can live up to its name remains to be seen, but the forces of resistance do seem to be on the rise (Read, 2019; see also Chapter 13 in this book).

As we have seen, Extinction Rebellion (or XR) has three principles or demands:

1. Government must tell the truth by declaring a climate and ecological emergency, working with other institutions to communicate the urgency for change;

2. Government must act now to halt biodiversity loss and reduce greenhouse gas emissions to net zero by 2025;

3. Government must create and be led by the decisions of a Citizens' Assembly on climate and ecological justice.

All these principles deserve critical consideration and ongoing debate (see Farrell et al., 2019), and reasonable people can accept them, challenge them, or disagree with aspects of them. Indeed, XR itself views these demands as part of an ongoing process of discussion and refinement, and how the movement and its key issues are framed has not been free from criticism, even by sympathetic voices (see e.g., Resilience, 2019). We have seen a good dose of critical reflection in this book too, not just unqualified praise. What is clear is that achieving the goals of XR will raise all sorts of deep complexities and thorny challenges, which may only be resolvable – if resolvable at all – through the messy process of lived experience and experimentation.

Nobody has all the answers; a swift decarbonisation of the global economy is an intimidating task, supported by the science but utterly unprecedented in human history; there is no detailed blueprint to tell us how to do it. But there is a clear distinction between XR and most other forms of thinking and practice in the environmental movement today. In the attempt to respond appropriately to climate breakdown and the broader environmental crisis (see Steffen et al., 2015), we have seen that XR is explicitly holding up nonviolent civil disobedience as an important and perhaps necessary part of the socio-political strategy for achieving a just and sustainable world (Extinction Rebellion, 2019a; Hallam, 2019).

In this closing essay I will argue that XR and rebellions like it are almost certainly going to grow in coming months and years as more people around the world become politically frustrated, angry, scared, and directly impacted by inaction in the face of today's overlapping ecological and humanitarian crises. I call this anticipated growth in XR and related movements the 'rebellion hypothesis', and I explain and defend the hypothesis below. Although I sympathise with the broad goals of XR, and have participated in many XR events, my argument herein is not that this and related movements *should* grow – a question I leave open for readers to determine for themselves. My argument is that they *will* grow, as behavioural shifts in society (or psychological tipping points) are provoked by the ongoing deterioration of Earth systems and rising existential threats to the community of life (Tollefson, 2019).

Put otherwise, I will argue that inaction has diminishing marginal returns, which makes social mobilisations for change more likely over time, since the real and perceived cost/benefit analysis of the environmental predicament tilts in favour of collective action. Whether this mobilisation occurs in time to avoid worst-case scenarios, however, is unknowable. Although my focus here is specifically on XR, the primary argument is about the rise of environmental activism more generally in coming years and decades, irrespective of whether these uprisings continue to march under the banner of 'Extinction Rebellion'.

In this post-script I will also assess the unsettling strategy of 'civil disobedience' – the practice of non-violently breaking the law to advance social, political, or environmental causes. Uncomfortable though it can make us feel, it is important for a society to understand the motivations for civil disobedience and evaluate the reasons given for practising this radical and disruptive strategy for societal change. Some commentators will be tempted to dismiss XR activists as mere 'trouble-makers' or even 'criminals', but such reactions, though understandable, risk mis-characterising these ethically motivated actions that are designed to be confronting, inconvenient, and disruptive – for a noble cause.

Even though most of us probably have reservations and concerns about civil disobedience, we must nevertheless appreciate that many of the most significant social and political advances over the last century owe much to social movements that engaged in civil disobedience as a primary strategy (Chenoweth and Stephen, 2010). One might think especially of Gandhi and the independence movement from British rule, the suffragette movement, and the civil rights movement. Rupert has rightly noted that the challenges of these movements and the challenges of XR are qualitatively different. Nevertheless, these esteemed traditions raise the disconcerting question: might future advances in society also demand civil disobedience?

DEEP HISTORY, DEEP FUTURE: AN ECOLOGICAL ACKNOWLEDGEMENT OF COUNTRY

Before looking more closely at XR, some context is required to fully understand this movement. Accordingly, I would like to begin this Australian-based essay, as one often begins a talk, by acknowledging the traditional custodians of the land on which I write – the Wurundjiri people of the Kulin Nation. I pay my respects to the elders, past, present, and emerging. These have always been lands where people have gathered for purposes of conversation, collaboration, and self-governance, and I feel honoured to be participating in that tradition, even though I find myself in the complex situation of occupying land whose sovereignty has never been ceded. I am still learning how to belong.

But what does it mean to acknowledge the traditional custodians of this (or any other colonised) land? It is very easy to say these things; it is much harder to know what it actually means; harder still to apply and live its truth. Let us briefly recall the colonial history in Australia to which I refer. In 1788 the British Crown turned up in Australia as a military force, and despite seeing the diverse cultures of Aboriginal Australians living on this land, the Crown declared *terra nullius*, which Australian readers will know translates as 'empty land' or 'land that belongs to no one'.

It was assumed that this land was empty because there was nothing which the Crown recognised as 'civilised people' living in Australia, despite the fact there was an Aboriginal population of somewhere between 300,000 and one million (Pascoe, 2018). Since the land was 'empty' according to these self-serving colonial assumptions, this gave a thin veneer of legitimacy to the occupation of Australia – an act of interpretive violence that of course soon evolved into acts of violence plain and simple. Indigenous populations do not often or ever freely give up their land or rights of self-governance to invading nations. Therefore, the stronger military powers have to resort to massacre and violence. Australian history is an example of a broader colonial history.

This colonial history, which still resonates in cultural and institutional reality today, is especially troubling in the context of the environment crisis we find ourselves in, so let me briefly dwell on this connection. Recent archeological evidence suggests that indigenous Australians have walked these lands for probably 65,000 years or longer (Pascoe, 2018). At once there is a striking lesson here: Australia's First Peoples did not undermine ecosystems in fatal ways. I do not want to romanticise indigenous culture or suggest that Aboriginal Australians did not have impacts on ecosystems and wildlife. They did. But the fact is that the First Peoples were able to live on this land for tens of thousands of years without degrading the land-base or fundamentally destabilising Earth systems. On the whole, ecosystems were able to regenerate sufficiently to allow for traditional cultures to be maintained over tens of thousands of years. It could be argued that this type of longevity or sustainability is the first and most important feature of any truly civilised culture: viability through deep history and capable of living on into the deep future. And yet, Aboriginal cultures were dismissed as uncivilised and primitive – invisible through the colonial lens adopted by the British Crown.

Compare this, then, with the industrial civilisation which the British Crown brought with it and established, which is merely two or three hundred years old. Over this very short time frame – a blink of the eye in geological timeframes – human beings have become so destructive that we have become geological forces. So significant has been our impact that Earth scientists now speak of the 'Anthropocene' – the first geological era caused by humans (Steffen et al., 2015). In fact, industrial civilisation is not so much an era as it is an event. Our industrial and extractivist form of life is decimating wildlife populations and driving ever-more species to extinction, deforesting the planet, destroying topsoil, disrupting the climate, emptying the oceans and poisoning waterways, overconsuming renewable resources, and is overly dependent on non-renewable resources. Plastic is contaminating essentially every ecosystem on Earth, from the deepest reaches of our oceans to the most distant corner of Antarctica. In the haunting words of James Lovelock (2010), the face of Gaia is vanishing.

So, we might fairly ask ourselves: which way of life, in the greater scheme of things, is more civilised? Is it the dominant culture and economic system today, which in a matter of a few centuries have degraded this rich ecosystem in ways that are threatening the viability of our species and all other species? Or is it the culture that was sufficiently civilised to live on the Australian continent for 65,000 years without destroying the planet?

I'm not going to suggest simplistically that we should try to return to the Aboriginal way of life, and it's quite possible that the land-base could not support today's Australian population of 25 million living off the land in that way. But I want to pay the most humble respect to the traditional cultures of Australia's First Peoples, for their ability to live for tens of thousands of years on this land, and to suggest that there will be features of indigenous ways of living, in Australia and elsewhere, that we have much to learn from, as we seek to respond appropriately to the range of deep environmental and social problems that modern, growth-orientated, industrial life presents. So it's not about a return to the past so much as it is about honouring the past and learning from it, as we move into a complex and turbulent future (Norberg-Hodge, 2009; Pascoe, 2018).

And a turbulent future it promises to be (Gilding, 2011). Sometimes scientists put bacteria in a Petri dish, on an organic substrate, and watch as the bacteria grow in numbers until the dominant colony has consumed all the available resources or poisoned itself from its own waste. In a sense, the bacteria grow themselves to death, like a cancer cell, by undermining the life support system upon which they depend, killing the host. But suppose we were aliens on Mars with a strong pair of binoculars and we were watching the happenings on Earth over the last couple of centuries. Could not industrial civilisation on Earth resemble the dominant colony of bacteria in the Petri dish? Are we not also at risk of consuming all the available resources and poisoning ourselves from our waste streams? It is a provocative metaphor but a useful one to get the analysis underway. And perhaps the background question that lies in the sub-text of this essay is this: can we, homo sapiens – so-called 'wise humans' – show ourselves to be smarter than common bacteria and avoid their fate?

In later sections of this essay I will reflect on the theory and practice of civil disobedience, both generally and in application to Extinction Rebellion. But first I'm going to ground my primary argument by presenting what I'm calling the 'rebellion hypothesis'.

THE REBELLION HYPOTHESIS: WHY A NEW WAVE OF ACTIVISM MAY BE COMING

It seems to me that there is a collective rumbling in the world today; a growing anger and anxiety about the troubled future that is unfolding day by day, and a growing sense that, if governments are not going to act decisively in response to today's overlapping ecological and social crises, then ordinary people like you and me will have to be the driving force for change. But feeling anger and anxiety about environmental breakdown and the unfolding extinction of species does not automatically translate into collective action. The history of widespread apathy or half-hearted resistance testifies to this truth. I know people who share their sense of dread with me but who have yet to mobilise and connect with activist groups. And I know a huge number of people who understand that the world is going to hell but who manage to distract themselves with modern engagements (Netflix, social media, etc.) in order to avoid facing the truth of our global predicament. We all see this social phenomenon which serves to entrench the status quo, and at times, I am sure, we all fall back into that default mode of apathy or inaction ourselves. It is easy to become disenchanted with the world and collapse into resignation or even despair (Bendell, 2018).

Why is it so easy to be complicit in ecocide and do little to resist? Even though the world is burning and billions of people are living in conditions of humiliating destitution, life for many of us in developed nations is relatively comfortable. Indeed, Australia is almost on top of the world in terms of prosperity, having more or less ducked the Global Financial Crisis ten years ago and our fossil-fuelled economy, prior to the Covid-19 disruption, at least, was growing at a robust pace. Although the disruption caused by the pandemic has cast many more people into economic insecurity, many still have discretionary income to spend on a new pair of jeans or shoes, or a new computer or device, and so forth. With important exceptions that must never be downplayed, not many people in the so-called 'first world' go hungry – even if, again, the unsettling impacts of the pandemic must be acknowledged.

And if our summer days reach infernal temperatures, we are generally able to turn on our air-conditioners and temporally hide ourselves from harsh ecological realities of global heating. The supermarket shelves are generally well stocked, there is petrol at the service stations, and on-demand streaming television is always waiting to sedate us if we need to self-medicate. Never in history has a comfortable and prosperous citizenry ignited a revolution or rebellion. When life is good, people do not mobilise to overthrow the system that seems to give them what they think they want. On what basis, then, do I formulate the rebellion hypothesis? Why should we expect mass mobilisation of people in coming years?

Let me explain by way of a simple allegory. Suppose you are on a boat, with a large cake, and you suddenly notice that the boat has sprung a leak. The leak is slow and you do not panic. Instead, you cut yourself a slice of cake and it is delicious. As you finish the slice, you assess the leak again. A little water has gathered in the bottom of the boat, but nothing too alarming. So you cut yourself another slice of cake. This slice was also delicious, but perhaps not quite as good as the first one. Upon finishing your second slice you notice that your feet are wet, which is a bit unsettling. It seems the leak has gotten worse and yet you wouldn't mind another slice of cake. What do you do? Is it time to panic and act? Or do you have another slice of cake?

The point of this simple story is to highlight how over time the costs of inaction can grow and the rewards of doing the same old thing begin to decline. This shift will eventually influence our behaviour. To borrow the language of economics, we might say that inaction has diminishing marginal returns. At first, inaction doesn't seem to cost much and might even offer rewards. Over time, the costs of inaction rise as the problems get worse. In this example, each piece of cake isn't quite as good as the last, while at the same time, the costs of not addressing the leak are becoming ever more pressing.

At some point – the tipping point – it becomes clear that the costs of inaction outweigh the benefits of more cake. At that point, the person in the boat switches from being passive consumer into an engaged activist (of whatever form). The meaning of their life has become animated by the desire to stop the boat from sinking. They have come to see that their life will be better if they act – and so they act. Cake is no longer as important or as desirable as stopping the leak. So, they substitute one form of life for a different form of life, because the calculus has changed regarding actual or perceived costs and benefits.

Those people already engaging in individual acts of resistance or collective action have passed their tipping points (Extinction Rebellion, 2019a). They have weighed up the costs and benefits of inaction and concluded that inaction now costs too much. At some point the conscious or semi-conscious calculations regarding the question of whether to rebel produced a positive answer – whether that moment was last week, last year, last decade, or, for the more seasoned activists, even last century. Calls for a 'new environmental radicalism' will be heard more loudly (Hamilton, 2011).

The notion of a tipping point is normally used in relation to ecological systems, where small increments in damage can suddenly lead to swift and drastic change, often irreversible. I am using the same idea but applying it to the human psychology of activism. Every day we become more aware that our planet is dying, putting the entire community of life at risk. Whether it is the Arctic or the Amazon burning, or a new species that has gone extinct, or a new climate report explaining why breakdown is happening faster than expected: each of these moments of awareness begin to add up, and yet often people don't respond with action or resistance. People can be bombarded with grim information about the ecological catastrophe unfolding, and yet remain locked in the ruts of life, doing today more or less what they did yesterday. Like the person in the boat, it is easier to continue eating or pursuing cake. This path is easily followed most of the time. But as Albert Camus (2000: 19) once wrote, 'one day a "why" arises – and everything begins in the moment of weariness tinged with amazement. "Begins" this is important.' One day a person asks: Are we the people we have been waiting for? If not us, then who? If not now, then when?

Perhaps my argument is getting clearer. I am suggesting that there is a growing 'affect' for resistance and rebellion in the world. When I speak of the rebellion hypothesis, what I am suggesting is that in coming months and years, more and more people will join XR or related movements as the costs of inaction continue to rise and the rewards of being a passive bystander decline. It seems to me that this is more or less inevitable because the costs of environmental damage will inevitably increase and become ever-more personal and immediate as capitalism continues to cannibalise itself and the planet. A new economics of activism is dawning. Currently, so much of the violence being imparted by our industrial civilisation is being externalised to other parts of the world or to others less fortunate or less powerful, including other species. This makes it easier to pretend that everything is fine and that we are not in an emergency. But as climate breakdown continues and the broader environmental crisis intensifies, the impacts will begin to be felt by more and more people, even in rich nations.

For example, when extended drought returns to (or intensifies in) Australia, as it seems destined to do, we will see more farmers joining XR or related movements as their livelihoods are directly threatened by climate change. Their tipping point will pass, and climate inaction as we know it today will be intolerable – a direct existential threat to their way of life (Fookes, 2019). When those droughts lead to increases in food prices, the urban consumer might stop to think: hang on, this had been predicted by the scientists and it has begun to affect me personally – I had better act. Their tipping point will pass. When the icecap disappears in coming years or decades, and the threat of rising seas levels becomes not a theoretical possibility but a practical problem, people living in coastal regions of the world will realise (as many already do) that climate change is not an abstract problem but something that could wash away their homes. Others will be affected by extreme weather events, and their tipping point will pass also. When children realise that they will be inheriting an unstable climate system, or a world without panda bears and the Great Barrier Reef, they will mobilise and agitate, and soon enough they will enter the voting constituency and provoke a profound political shift. Their tipping point will pass. One could go on.

Of course, to some extent this growing resistance is already underway – XR is hardly the first mobilisation in this vein. Think especially of the noble work of the global School Strikes or, in Australia, the anti-Adani activists resisting new coalmines. The list is long, diverse, and esteemed. But my argument is that the costs of inaction are necessarily going to increase, and as the Earth system deteriorates, the benefits of passive by-standing are going to seem less and less rewarding and socially acceptable. In other words, ever more people will experience a tipping point. Your neighbour, your colleague at work, a child or police officer, perhaps eventually more politicians. Each of them has a threshold or tolerance – and their tipping points are approaching. A cultural shift may be underway, even if it remains in its early stages. This cultural shift could eventually filter upwards and have political and macroeconomic effects. (One must also accept that this social energy at times might be misdirected in regressive ways as people look for minority scapegoats to blame for the harder economic times – a complex issue that is noted but deferred for analysis on another occasion.)

We are all in a lifeboat called Earth. In the 1960s and 70s when the modern environment got underway, people noticed a leak in the boat and recognised it to be dangerous (Meadows et al., 1972). They spoke of a crisis in the future. Things continued to get worse but most people couldn't resist the cake. Not enough people mobilised to plug the leak. Now the boat is leaking disastrously, and water is up to our necks; some people are already drowning. Crisis has arrived. The future is now. And more and more of us are sick of cake. More and more of us have exceeded our threshold. To change the metaphor, the floodgates are threatening to burst and it is not clear that the growing energy of opposition can be contained (Extinction Rebellion, 2019b).

Furthermore, this growing force is going to lead to increasing pressure within society – like steam increasing in a closed system. As the resistance increases and becomes more energised, we can expect backlash from those still benefiting from the existing system (and again, we are seeing this already). But as the defenders of the status quo lash out and oppress the rising tide of resistance, what they will discover is that their actions in fact only mobilise more people, as the social license of the fossil industry, corporate greed, and the politics of denial fade and ultimately disappear. In other words, one day state and corporate blindness or apathy in the face of worsening ecological catastrophes will offend public morality, and perhaps that day is closer than we think.

Based on empirical studies, it has been estimated that only 3.5% of a population needs to mobilise and engage in collective action to induce deep structural and cultural change (Chenoweth, 2017; see also chapters 6 and 13 in this book, and the Appendix, for further discussion). While such estimations need to be interpreted critically and cautiously – and every context and situation is different – the point is that surprisingly small social mobilisations can have far-reaching impacts. Of course, such deep transformations do not happen overnight, but the history of disruptive social movements shows that things can happen faster than one might at first think. The Environmental Movement may not need a Martin Luther King or a Gandhi to lead. Perhaps what is needed is a thousand or a million Rosa Parks to get things done.

Due to the momentum of global capitalism today, global environmental problems are almost certainly going to get worse before they get better, and this will only fuel the fire of rebellion. There is an ecological contradiction built into our society, our economy, and our politics: that contradiction is the assumption that limitless economic growth is possible on a finite planet (Hickel and Kallis, 2019). But even the simplest of folks can grasp that when something cannot continue, it stops. We no longer need to ask, 'can we change the world?' – because the world is inevitably going to change and is already changing. The future is not what it used to be. One way or another, change is coming because the status quo simply cannot be maintained (Read and Alexander, 2019). We are in the process of witnessing a self-destructive civilisation collide with environmental limits, and increasingly people are going to suffer under this perverse system, and increasingly people are going to see that better, freer, less impactful, and more compassionate ways of living are available. People will try to live those new worlds into existence. Both of these things – both suffering under the existing system and the prefigurative 'new world' imagination – are mobilising forces.

My prediction or hypothesis, then, is that this collective rumbling – this emerging matrix of global social movements (Read, 2019) – is only going to intensify and amplify. At some point, it may ignite in ways that currently our imaginations cannot even begin to grasp. Or it may fade away into oblivion like other beacons of hope – think Occupy, for example, which rose as quickly as it fell (even if we can still debate whether Occupy induced valuable impacts and conversations that live on). Social movements have a tendency to surprise us. I am not sure whether forthcoming environmental rebellions will be able to *save* the world, but I feel they are destined to *change* the world as the world changes us.

WHAT IS CIVIL DISOBEDIENCE AND IS IT JUSTIFIED?

Let me now spend some time examining a defining feature of XR – that is, an openness to civil disobedience as a strategy for change. What is civil disobedience? And when, if ever, can it be justified?

In essence, civil disobedience can be defined as 'a public, nonviolent and conscientious breach of law undertaken with the aim of bringing about change in laws or government policies' (Brownlee, 2007: np). For present purposes I will assume that for disobedience to be 'civil' it has to be nonviolent, and indeed this accords with the explicit and unconditional commitment XR has to nonviolence (Farrell et al., 2019). In an important aside, empirical studies show that movements committed to nonviolent disobedience tend to be twice as successful in achieving their aims as violent demonstrations (Chenoweth and Stephan, 2012), thus XR's principled commitment to nonviolence is also pragmatic and evidence based. One might add that it is also a diverse strategy – Gene Sharp famously listed 198 ways to practise nonviolent resistance (Sharp, 1973). Before engaging in such acts, however, individuals and groups should ask themselves: can civil disobedience ever be justified in a democracy?

It can be helpful to begin assessing civil disobedience in relation to basic democratic theory. Imperfect though it is, it can be said that we, in Australia, live in a democracy. Many readers, I suspect, will also be living in societies with a democratic self-image. Among other things, this means that citizens and permanent residents get to vote on who will represent them in government, and government includes a legislative branch that creates law and an executive branch that enforces it. (For present purposes I'll leave to one side the judicial branch that interprets law – or rather, creates law through its interpretations.) Since we all have, in theory, an equal opportunity to influence the law-making process through the ballot box, it is generally assumed that we should obey the law because the democratic process is the best way to organise and structure society and develop public policy that serves the common good.

From this perspective, an opponent of civil disobedience might argue as follows: *We can't all break the law every time we disagree with it. Imagine how unstable society would be if that happened. If we don't like what is happening, we can campaign for change like everyone else, and if we succeed, we can vote the existing government out of power through the electoral process and vote in a new government.* In this way, democratic societies are said to have created the institutions and processes needed for their own peaceful improvement. It may not be a perfect political system, but as Winston Churchill is reported to have said: 'It is the worst form of government, except for all the others.'

So, the main objection to civil disobedience is this: *If you disagree with a law or policy, don't break that law or policy; instead, campaign to get it changed through the democratic process. If you are permitted to break the law just because you disagree with it, then why can't anyone break a law they disagree with?* At first instance, perhaps, this objection seems quite powerful. Indeed, the great philosopher Immanuel Kant argued that '[a]ll resistance against the supreme legislative power… is the greatest and most punishable crime in the commonwealth, for it destroys its very foundations' (Kant, 1970: 81). If people only abide by laws they agree with, then the rule of law would break down. To some extent, then, we might all have sympathy with the political assumption that we ought to obey laws – even laws we don't agree with.

But it is one thing to make that broad and pragmatic concession. It is quite another to suggest that all laws, always, ought to be obeyed. If obedience to law were unconditional and absolute by virtue of the democratic process, it would follow that civil disobedience is always unjustified. How might acts of civil disobedience be interpreted within the contested disciplines of legal and political theory?

First of all, one might argue that civil disobedience is potentially justifiable when the mechanisms of democracy are not working properly, that such laws do not represent the will of the people. This can occur when laws and policies are shaped by the undemocratic influence of foreign governments, billionaires, mass media conglomerates, or other corporate lobby groups (e.g., buying a politician's support) (see, e.g., Mayer, 2016; Tham, 2010). In such cases, one might suggest that laws produced by undemocratic processes do not demand our political allegiance since they were not produced through fair, robust, and representative democratic processes.

There is also a second way in which it might be argued that civil disobedience is justified. That is, to recognise that there is a distinction between law and morality; or a distinction between what is law and what is just. Often, we might admit, there is much overlap between law and justice. The more overlap the better. But any thinking person knows that often in history, and no doubt still today, there are times when we see a clear difference between what is 'law' and what is 'just' – even if justice is an essentially contested term. In other words, democracy may be the best form of government, but this does not mean that a democracy always gets things right. Rather, democracy, when it is functioning properly, reflects culture, and there is *no reason to think that cultural norms and expectations are always just*. Put more directly, a functioning democracy can produce unjust laws when a citizenry knowingly and voluntarily votes for policies that are unjust (even if they are not considered unjust by those voting for them).

For example, we know that democracies have historically declared it illegal to engage in same-sex relationships, and today most members of liberal democracies recognise that such laws were and are in breach of basic civil rights. In the past, laws produced in democracies have institutionalised slavery, ratified unjust wars, legally entrenched racial segregation, criminalised homosexuality or particular religious practices, prohibited women and people of colour from voting, and so forth. Again, what is law does not automatically overlap with what is just. Nobody can deny that unless they still believe in the 'Divine Right of Kings' – and I am sure no one thinks that today's world leaders are God's infallible messengers on Earth chosen to lead us to the Promised Land.

At such times when a law or policy is clearly unjust (e.g., recognising ownership of persons as slaves), a case can be made that there is a place for civil disobedience in democratic societies, on the grounds that we must accept that even democratically produced laws sometimes get it wrong – sometimes *really wrong*. There is a rich and revered tradition in legal and political theory that recognises and accepts these broad lines of argument (see review in Brownlee, 2007). In other words, it is widely accepted that there is a proper place for civil disobedience in liberal democratic societies. In fact, as we look back on social movements in history – whether it is Gandhi's campaign for independence, Martin Luther King, Jnr and the civil rights movement, or Emmaline Pankhurst and the suffragettes – some of the greatest leaps forwards in social and political progress have been a *result* of acts of civil disobedience. It would show a gross lack of historical understanding to dismiss civil disobedience as a regressive social practice. The powerful but uncomfortable inference is that future acts of civil disobedience may also be required to advance our state of society.

CIVIL DISOBEDIENCE AND EXTINCTION REBELLION

So how does this apply to Extinction Rebellion? There are, as I have just implied, two main ways to evaluate civil disobedience. On the one hand, an argument could be made that we live in democracies that are at least partially broken, such that the laws and policies that are produced are sometimes undemocratic because of the undue influence corporate interests have had on the legislative process – for example, the fossil fuel industry, the Murdoch media, or other powerful economic forces (see, e.g., Market Forces, 2019; Tham, 2010; Cooke, 2019; Knaus, 2018; Rudd, 2019). This suggests that even if our culture wanted a strong climate response, vested interests would interfere with any such response and ensure that law and policy kept things more or less as they are. To some extent, this may be part of the reason why climate policy around the world is often weak and sometimes non-existent. In cynical words often attributed to Emma Goldman: 'If voting changed anything, it would be made illegal.' One might say in the same vein: if lunatics have taken over the asylum, a case can be made for the citizenry to break their rules and establish new ones.

Perhaps the more powerful argument for civil disobedience, however, is that, overall, dominant cultures today have yet to fully appreciate the magnitude of climate breakdown and the broader environmental crisis (perhaps due to powerful vested interests shaping public consciousness). After all, as noted earlier, it is still quite easy to distance ourselves from the impacts of these crises, and we also know that Australia, for example, has a government that celebrates coal and essentially denies that climate breakdown is a problem deserving of a significant response. For these reasons among others, the Australian government is each instant losing some of its integrity. And Australia is not alone.

So, we might draw an analogy here with the anti-slavery or civil rights movements in the US. Where once the state sanctioned and supported the moral wrongs of slavery and segregation, today the state sanctions and supports the moral wrong of climate breakdown. Activists who engaged in civil disobedience during the civil rights movement might accept that white people were in fact voting for racist laws and public policy, but justify their disobedience on the grounds that racist laws and policies were wrong and deserved to be disobeyed. We cannot say that the anti-slavery activists or civil rights activists were wrong to break the law and engage in nonviolent acts of civil disobedience. Those racist laws were grossly immoral, and they deserved to be disobeyed. Rosa

Parks was right not to give up her seat on the bus on that fateful day in 1955 even though it violated the laws and regulations. According to Henry Thoreau (1982), who published his famous essay on civil disobedience in 1849, this strategy is not just a right but at times a duty. It is no surprise, then, that Gandhi, Martin Luther King, Jnr, Emmaline Pankhurst, and countless other social activists have been inspired to engage in such acts and are now revered for their bravery.

Let us ask with Thoreau: are we expected to resign our conscience to the legislator? Why have a conscience, if we are simply expected to uncritically affirm all acts of government? We must be human beings first and subjects of the state afterward. As Thoreau (1982: 111) argued, 'it is not desirable to cultivate a respect for law, so much as for the right', and indeed, he insisted that respect for law can, at times, make us daily agents of injustice. In relation to his own time, Thoreau argued that one could not be associated with the US government without disgrace, for he could not recognise as *his* government what was also the *slave's* government. He concluded that if a government's law is of such a nature that it requires you to be the agent of injustice to another, then, he argued: break the law. 'Let your life be a counter-friction to stop the machine,' he declared (Thoreau, 1982: 120). 'Cast your whole vote, not a strip of paper merely, but your whole influence' (Thoreau, 1982: 122).

Since its emergence, thousands of XR activists have been arrested for engaging in civil disobedience. While no one should fetishise 'being arrested' as the only way to participate in XR, and the movement should recognise also that people have different 'biographical availabilities' for being arrested (Beyerlein & Bergstrand, 2013), the fact is that all acts of civil disobedience raise the possibility of being arrested and perhaps imprisoned. No doubt acts of civil disobedience will be perceived by many as annoying and inconvenient and unnecessarily disruptive, but that calculus always has to be weighed against the moral wrong that is motivating the disobedience (see Monbiot, 2019). Slavery and segregation were also 'inconvenient'… for those who suffered under racist laws.

In that light, the inconvenience caused by 'sit-ins' and bus boycotts pale in comparison. Similarly, when environmentalists engage in acts of civil disobedience to resist ecocide, the extinction of species, and the unfolding climate emergency, some sectors of society will no doubt be appalled and dismiss the activists as common law-breakers or radical anarchists. Civil disobedience may indeed be inconvenient to many people. But to evaluate the legitimacy of the civil disobedience, one has to resist superficial analyses and ask how that inconvenience compares to the future suffering, and indeed the suffering already being caused, by environmental breakdown (Spratt and Dunlop, 2019; Nixon, 2013).

I can now bring the analysis to a head. Just imagine, for example, that in ten years or twenty years or thirty years – it doesn't really matter when – we discover that our high-impact modes of production and consumption have led to even more alarming ecosystemic breakdown, a future that has mountains of scientific support (see, e.g., Steffen et al., 2015; Spratt and Dunlop, 2017). Suppose the climate reaches its tipping point; Australia and other nations enter indeterminate and intensifying drought (just look at New South Wales or the bushfires of 2019/2020); food production drops even as population grows, leading to mass famine and increased geopolitical tension and war; suppose in ten or twenty years the Arctic icecap disappears and the methane release from the permafrost induces a swift jump in global temperatures. Suppose any number of such things happen and people begin to die in greater numbers. When we look back on today, we will ask ourselves: Did we do enough? Were we complicit in a broken system? Should we have been so obedient given that we knew our gutless governments were leading us down a dead end?

These questions are not for me to answer – I am still struggling with them myself. I will remain a sympathetic critic and revise my views as new evidence and insight emerges. None of us can condemn or condone the actions of XR in advance of their particular, context-dependent manifestations. One might sympathise with XR in general while disagreeing with specifics, or vice versa. These are very personal questions (with social effects) which we must meditate on with due diligence.

But my point is that if the future turns out how the best scientists are predicting it will turn out if business as usual continues (for reviews, see Steffen et al., 2015; Spratt and Dunlop, 2017; Spratt and Dunlop, 2019), then the younger generation might well ask us what we did to resist the foreseeable collapse of ecosystems and the humanitarian catastrophes such breakdowns will induce (and are already inducing).

CONCLUSION: ON THE RIGHT SIDE OF HISTORY?

Writing in the 19th century, Karl Marx announced that he had discovered the laws of history. He maintained that it was inevitable that as the contradictions of capitalism became ever more severe and transparent, eventually the working class would rise up and overthrow the capitalist class and establish communism. I have always been suspicious of determinist conceptions of history, knowing that human societies do not follow predetermined laws. I feel that we will be what we make of ourselves and nothing else, as the existentialists argued, even if we are born into a world not of our own making. But when we freely act in ways that undermine the ecosystems that we (and future generations) depend on for freedom and prosperity, then our lives begin to be shaped not so much by human decisions as by ecological realities and geological forces. At least, human freedom is increasingly *contained and influenced* by those worsening realities and forces. We are living in such times today.

What Marx never foresaw was that capitalism would indeed fall, but not by way of revolution, but by way of deterioration and perhaps collapse. As the broad ecological crisis intensifies, and collapse situations become more common, challenging, and disruptive, I have argued that more and more people will face their psychological tipping points and become engaged in collective action. At some point, tolerance of ecocide will become intolerable.

What is the threshold of your neighbour? Your children or parents? Your work colleagues? Our politicians? I don't know, but my sense is that those tipping points are approaching – if not tomorrow, then next month, or next year, or the year after that. The rebellion hypothesis, as presented, is that every day more people are saying to themselves: 'I am an activist; I am a change-maker not a passive consumer; I am responsible for participating in progressive social change; I want to be; I have to be.' The question we must all face, as global citizens on a dying planet, is whether our governments are meeting their fundamental duty to keep us, our children, and the broader community of life, safe.

Rebellion, I am suggesting, has effectively become a law of history due to ecological realities. The climate crisis is already here, to some extent locked in, and certainly threatening to get much worse. Three hundred years of industrial momentum means that it is now too late for any smooth, non-disruptive democratic shift to some ecological civilisation. For better or for worse, turbulence and disruption will define coming decades. Things are likely to get worse before they get better. But as this happens people will inevitably be mobilised as the calculus of apathy and inaction shifts and the activist is born. As Camus declared: 'Everything begins in that moment of weariness tinged with amazement.'

In six months, or two years, or five years, or ten years, I invite you to reflect back on this essay and assess to what extent you think the hypothesis presented has been verified by growing global social movements or falsified by increased apathy. I feel confident, for the reasons I have outlined, that the future will confirm my prediction. If I am wrong, and the status quo endures, then all the worse for us. In short, I feel the logic of rebellion is becoming irresistible to more and more people and that this trend is destined to continue. And one implication of this is that we should not conceive of XR as something already riding the crest of a wave, but rather, XR represents a movement of movements that is still in its infancy. This entails a prospect of something much bigger that is still in the process of being born, even though it may be that the hour is darkest just before dawn. Still, the promise of a new dawn is not needed to justify the rejection of a world immiserated by capitalism's violent overreach.

In closing, I have invited you herein to ask yourself: what are the costs and benefits of inaction? What are the costs and benefits of resistance and rebellion? What are the rewards of building a new world arm in arm with your neighbour? My argument has been that this calculus is already shifting in favour of resistance, rebellion, and renewal, and that this shift is now unstoppable, whether one sympathises with XR or not. XR is part of this shift but the global movement and energy are broader than any one framing or articulation. The floodgates are holding for the time being, but the laws of physics will win out, as they always do. Participants in XR are early adopters and if warnings of ecological science are to be taken seriously and prove even vaguely accurate, this movement – despite the negative press it will inevitably receive from some sectors in society – is likely, as George Monbiot (2019) argues, to end up on the right side of history.

With a nod to Thoreau, Extinction Rebellion seeks to be a counter-friction to the machine.

POST-SCRIPT REFERENCES

Bendell, J., 2018, 'Deep Adaptation: A Map for Navigating Climate Tragedy', IFLAS Occasional Paper 2 (27 July 2019).

Beyerlein, K., & Bergstrand, K., 2013, 'Biographical Availability', *The Wiley-Blackwell Encyclopedia of Social and Political Movements*, viewed 5 September 2019, https://onlinelibrary.wiley.com/doi/abs/10.1002/9780470674871.wbespm012

Brownlee, K., 'Civil Disobedience', *Stanford Encyclopedia of Philosophy*, viewed 5 September 2019, https://plato.stanford.edu/entries/civil-disobedience/

Brulle, R., 2014 'Institutionalizing Delay: Foundation Funding and the Creation of US Climate Change Counter-movement Organizations', *Climate Change*, vol. 122, no. 4, pp. 681–694.

Camus, A., 2000, *The Myth of Sisyphus*, Penguin, London.

Chenoweth, E., 2017, 'It May Only Take 3.5% of the Population to Topple a Dictator – With Civil Resistance', *The Guardian*, 2 February 2017, viewed 5 September 2019.

Chenoweth, E., & Stephan, M., 2012, *Why Civil Resistance Works: The Strategic Logic of Non-Violent Conflict*, Columbia University Press, New York.

Cooke, R., 2019, 'News Corp: Democracy's Greatest Threat', *The Monthly* (May 2019).

Extinction Rebellion, 2019a, Official website, viewed 5 September 2019, http://rebellion.earth

Extinction Rebellion, 2019b, 'A Movement of Movements for the October Rebellion', viewed 5 September 2019, https://rebellion.earth/2019/08/29/introducing-the-movement-of-movements-for-the-october-rebellion/

Farrell, C., Green, A., Knights, S., and Skeaping, W. (eds), 2019, *This is Not a Drill: An Extinction Rebellion Handbook*, Penguin, London.

Fookes, T., 2019, 'Fear of Drought, Flood, and Fires Leads Farmers to Plea for Urgent Action on Climate Change', ABC News, viewed 13 September 2019, https://www.abc.net.au/news/rural/2019-09-13/fear-of-drought-flood-and-fires-leads-farmers-to-plea-for-action/11508834

Gilding, P., 2011, *The Great Disruption: How the Climate Crisis will Transform the Global Economy*, Bloomsbury, London.

Hallam, R., 2019, 'Common Sense for the 21st Century: Only Non-Violent Rebellion Can Now Stop Climate Breakdown and Social Collapse', pre-publication working draft, version 0.3.

Hamilton, C., 2011, 'We Need a New Environmental Radicalism', *Green Left Weekly*, 29 April 2011, issue 878.

Hickel, J. & Kallis, G., 2019, 'Is Green Growth Possible?', *New Political Economy* (in press). DOI: 10.1080/13563467.2019.1598964

Kant, I., 1970, *Kant's Political Writings*, edited by Reiss, H., Cambridge University Press, Cambridge.

Knaus, C., 2018, 'Australia's Political Parties Got $62m in "Dark Money" Donations Last Year', *The Guardian* (3 September 2018).

Lovelock, J., 2010, *The Vanishing Face of Gaia*, Basic Books, New York.

Market Forces, 2019, 'Friends in High Places: Fossil Fuel Political Donations', *Market Forces* (February 2019), viewed 7 September 2019, https://www.marketforces.org.au/politicaldonations2019/

Mayer, J., 2016, *Dark Money: How a Secretive Group of Billionaires is Trying to Buy Political Control in the US*, Scribe, London.

Meadows, D., Meadows, D., Randers, J., & Behrens, W., 1972, *Limits to Growth*, Signet, New York.

Monbiot, G, 2019, 'History Will Be Kind to Heathrow Climate Protesters Who Stop Us Flying', *The Guardian*, 4 September 2019.

Nixon, 2013, *Slow Violence and the Environmentalism of the Poor*, Harvard University Press, Cambridge.

Norberg-Hodge, H., 2009, *Ancient Futures*, Sierra Club, San Francisco.

Pascoe, B., 2018, *Dark Emu: Aboriginal Australia and the Birth of Agriculture*, Magabala books, Broome.

Resilience, 2019, Articles on Extinction Rebellion, viewed 5 September 2019, https://www.resilience.org/?type=Filter+by&s=%22Extinction+Rebellion%22

Rawls, J., 1971, *A Theory of Justice*, Belknap Press, Boston.

Read, R., 2019, 'How a Movement of Movements Can Win: Taking XR to the Next Level', *Rupert Read's Website*, 13 August 2019, viewed 5 September 2019, https://rupertread.net/writings/2019/how-movement-movements-can-win-taking-xr-next-level

Read, R. & Alexander, S., 2019, *This Civilisation is Finished: Conversations on the End Of Empire and What Lies Beyond*, Simplicity Institute, Melbourne.

Rudd, K., 2019, 'Democracy Overboard: Rupert Murdoch's Long War on Australian Politics', *The Guardian* (7 September 2019).

Spratt, D. & Dunlop, I., 2017, 'What Lies Beneath: The Scientific Understatement of Climate Risks', *Breakthrough Institute*, September 2017.

Spratt, D. & Dunlop, I., 2019, 'Australia's Climate Stance is Inflicting Criminal Damage on Humanity', *The Guardian*, 3 August 2019.

Sharp, G. 1973, '198 Methods of Non-violent Resistance', viewed 5 September 2019, https://www.aeinstein.org/wp-content/uploads/2014/12/198-Methods.pdf

Steffan, W., et al. 2015, 'The Trajectory of the Anthropocene: The Great Acceleration', *The Anthropocene Review*, vol. 2, no. 1, pp. 81–98.

Tham, J-C., 2010, *Money and Politics: The Democracy We Can't Afford*, UNSW Press, Sydney.

Thoreau, H., 1982[1849] 'Civil Disobedience' in Bode, C. (ed.), *The Portable Thoreau*, Penguin, New York, pp. 109–137.

Tollefson, J., 2019, 'Humans are Driving One Million Species to Extinction', *Nature*, vol. 569, p. 171.

APPENDIX

RUSHING THE EMERGENCY, RUSHING THE REBELLION?

This pamphlet was co-authored with two of my closest colleagues in XR, Marc Lopatin and Skeena Rathor. In a way it is a kind of sequel to my own pamphlet from 2019, 'Truth and its consequences'. In another way though it is an exciting new departure, urging as it does that XR needs to tell a new story by way of its actions (and words), a story that focuses upon our vulnerability to existential threats. With coaching from Marc, I gave this new story a trial-run during the October 2019 Rebellion, most notably when I appeared as the XR rep on the BBC's flagship 'QuestionTime' programme.[90]

The three of us were convinced by the end of the October Rebellion that, without a recentering around a new story along something like these lines, XR would not flourish, would not grow further, would certainly not realise 3.5% of the population as more or less active supporters, let alone the larger percentage that we actually need in at least passive support, for system-change. So we wrote this pamphlet, which appeared in January 2020. It is republished here as an Appendix to the book mainly because it is somewhat 'technical':[91] *this pamphlet was not intended for as wide a potential audience as most of the pieces in the body of the book. However,*

90 You can watch the key segment of the programme here: https://www.youtube.com/watch?v=QK7DKiKh9_Q

91 If you prefer an entirely non-technical presentation of the themes of the pamphlet, then watch https://vimeo.com/389093326

it will be of interest to those wishing to dive deeper into the 'theory of change'.

In a remarkable coincidence of timing, the pamphlet appeared at the very moment that the novel coronavirus appeared. As I've discussed above, especially in Chapter 26, the crisis that the latter rapidly generated had an effect curiously akin to that which we were aiming in our pamphlet to discuss – and to evoke. It was and is a worldwide crisis of felt vulnerability, requiring a vision in response to it. Thus the pamphlet is, I would argue, simultaneously in one way rendered unnecessary by history and in another way more pertinent and prescient than ever. Some of the pamphlet reads eerily almost as if we were aware that a scary new virus was being unleashed by the recklessness of eco-destructive economic globalisation at the very moment that we were writing, and its recommendation of making vivid our vulnerability by making realer our equality is also tremendously pertinent to where we find ourselves, as a virus-struck planet with an economy that we've protectively contracted, in 2020…[92]

CALLING ALL REBELS!

Over the last year and a half, you have come together and created something truly beautiful. You've put your heart and soul into it, and you've become a part of it. At times it has been stressful, painful, and exhausting. But it has also been extraordinary and deeply moving, and collectively, as part of the climate movement, you have brought the climate and ecological emergency into conversations in homes across the globe. As a result, Extinction Rebellion (XR) is one of the top influencers globally when it comes to the climate and ecological emergency.

But right now, XR is at a turning point. What happens next is its most important decision to date. It could propel us all to new heights in 2020, or see us plagued by the incoherence that many of you have been feeling since last October's Rebellion.

92 The pamphlet consists of personally held views of the authors and not those of Extinction Rebellion. We give our dear thanks to the following folks for helping develop and improve the pamphlet: Dario Kenner, Sarah Kingdom Nicolls, Jessica Brinton, Robert Possnett, Janie Skuse, Rei Takver, Pete Williams, Adam Woodhall and XR's Vision Sensing and Regenerative Culture teams.

This fork in the road is why the three of us agreed to write a pamphlet, albeit from the centre of XR. We apologise if you feel bombarded with information just now. Rest assured, we've done our best to make our key points in a short executive summary. We are keenly aware that some of what we put forward will make for uncomfortable reading. This is our gentle act of rebellion, because we believe XR is in a rush, and at risk of forgoing an untapped opportunity to grow the movement and inspire others outside of our numbers. We believe XR is rushing the 'emergency', which could belittle efforts to create the changes we desperately need at local, national, and global levels.

So this pamphlet is about how XR can grow and catalyse by first taking an honest and searching look at itself. What are our blind spots, tensions, and paradoxes that have produced success and incoherence in almost equal measure these past months? How can these issues be resolved so that XR can truly embody and embed the change it says it wants to be? These questions are why we absolutely encourage you to feed into the strategy consultation process that is ongoing right now. Your voice counts. And if you don't believe this, we will almost certainly fail as a movement.

Even if you don't read a further word of this pamphlet, please respond to the strategy team that is canvassing local and regional co-ordinators. Please share your vision for how XR might think the unthinkable and achieve it. If, like the three of us, you believe XR needs to be bolder, cleverer, more creative, and shapeshifting, then only good will come.

EXECUTIVE SUMMARY

This pamphlet is an invitation for rebels to coalesce around a new story and vision, which we hope can lead to embodied actions. That story is human-centric, as opposed to environmental. It is nearer-term (as opposed to far-flung), as evidenced by the vulnerability of civilisation to locked-in unpredictable and extreme weather. We tend to forget that we live in a world that has evolved as if the Climate and Ecological Emergency (CEE) didn't exist.

A ramification of this is our story untold. It is about how vulnerable the complex human systems that sustain our lives are to a near-term future characterised by shock, disruption, and even breakdown.

At the core of the new story is a vision for overcoming this vulnerability, by displacing inequality at the local, national, and global level. It is informed by the truth that no community, country, or continent will be an island, to paraphrase John Donne. Despite what some billionaires believe, none of them will be looking down on the rest of us from a space station anytime soon. It is, above all, a story about how Mother Nature is making us all one.

MOVEMENT BUILDING IN 2020

This pamphlet asserts that telling this story could be XR's central means of both movement building in 2020, and the wider catalysation of fellow travellers who don't yet consider themselves rebels. What follows, therefore, is intended as a departure from recent high-profile and overly polarising tactics such as XR co-founder Roger Hallam's leveraging of the Holocaust with mainstream media, and the Canning Town tube action of last October.

ACTIONS SPEAK LOUDER THAN WORDS

A central theme of this pamphlet is that XR needs to not only tell our vulnerability story but embody it through actions. In 2020, XR needs to include regenerative and restorative action so that our lived reality can pivot from vulnerability to radical equality. Actions that unfurl a vision for the future will tell a story of regeneration, rewilding, and repair. It means thinking more carefully and creatively about how actions might carry both the story of vulnerability and a vision for equality. Given such actions could be XR's primary means for movement building, creativity will be a must – alongside a broader invitation for participation. Within this resides the untapped opportunity to centrally frame actions with hundreds of decentralised affinity groups to create shape-shifting and nationwide impact. Inherent here is a shift in energy away from only regarding ourselves – or being typecast as activists driven by 'impossible demands'. It suggests finding a more gentle or nuanced energy essential for movement building and systems transformation.

THE SPIRIT OF INQUIRY

The pamphlet therefore challenges XR's existing 'theory of change' (ToC), not least around its guiding premise that we need only mobilise 3.5% of the population. That in itself creates tension with connecting and building regenerative cultures, internally and externally, through our messaging, storytelling, and actions. In response, XR needs the courage to honestly assess and evolve (not overthrow) its ToC. The April 2019 phase of Rebellion worked brilliantly but that may only take the movement so far. In the absence of a more compelling story and broader invitation to act, planning 'one more heave' on London's streets is a grave threat to the movement, and to the purpose to which XR is in service. At the present time, the general public is unlikely to perceive XR's return to the capital as anything other than a source of big irritation. Participating rebels risk being no different to climate scientists clutching the findings of their latest models, earnestly telling one another that this time the government must surely listen.

FAST TRACK TO COHERENCE

With the above in mind, we assert that embodying the full breadth of a new story could bring some urgent coherence to the unaddressed tension at the heart of XR. That tension being the observable conflict between XR's principles and values that speak to mobilisation on the one hand, and regenerative and inclusive cultures on the other. As this pamphlet will hopefully make plain, we seek to embody a vision of a brave and beautiful world.

SECTION 1: WHAT'S THE STORY?

We live and die by the stories we tell each other – and that story on the streets of London is changing.

– *Charlotte Du Cann, writing about the October Rebellion for* The New York Times

The story might be changing but not nearly fast enough. This is despite XR achieving extraordinary first-year growth and external impact. Extraordinary, because XR was saddled with the burden of a decades-old story about climate change that spectacularly failed to cut through and trigger an urgent planetary response.

Telling the story handed down by climate scientists has been like driving with the handbrake on. As a species, we're not wired to respond to slow-moving distant/invisible threats timed to deliver Armageddon decades into the future. It is why competing chatter about emissions reduction pathways has failed to activate anywhere near enough people worldwide to reach a tipping point response. In its place, GHG emissions have risen by 1.5% a year for the past decade, according to the UN Environment Programme annual emissions gap report for 2019.

We do not dismiss that many thousands of people – perhaps yourself included – have been activated by the climate change story as told. We have only to look at Fridays For the Future or our own Rebellion. You are in rare company as these numbers are still small and show no sign of going exponential. Greta Thunberg herself has been honest enough to say the school strikes have achieved nothing.

As the authors of this pamphlet, we go as far as to say that retelling the climate change story is essential for credibly framing XR's domestic (UK) and international targets for global net zero emissions. Not least because the existing story about the CEE is rooted in incremental, long-term temperature rise, which has failed to elicit anything like the mass consent required for a radical reduction of emissions.

This section is therefore an invitation for rebels to coalesce around a new external story of our vulnerability. Above all, the aim of the new story is to inspire movement building as a precursor for realising XR UK's strategy for accelerating the coming of system change.

DISRUPTING THE STORY

Many people – inside and outside the movement – are now wondering what happens next. Not least because our theory of change or strategy is about trying to bring about system change – as embodied by our second demand – by pivoting off a story about the future that's been failing for decades.

The story about climate change, as told, describes the wrong kind of emergency on just about every measure. And yet climate scientists and civil society have expected it, time and again, to peacefully deliver on the following:*

Over 1 billion people living across the minority world must accept a new normal for living.

Over six billion people living across the majority world must make peace with an aspiration to one day live like those in the minority world.

**The two statements manifest a crude demarcation and omit acute levels of inequality and social exclusion throughout the minority world.*

Spanning both statements is the realisation that the climatic forces that human activity helped set in train will not discriminate. Good luck to any of the super-rich who think digging a luxury bunker in the New Zealand wilderness is going to give them and their children a free pass. At last look, they were still trying to figure out how to avoid being murdered by their security detail, before runaway temperature rise wipes them all out anyway.

In the shorter term, there will be chaos and misery on an unimaginable scale as the most vulnerable people on the planet will not dutifully stay put and wither. They will seek security for their loved ones as we all will. The global food system for example already fails to properly nourish billions of people and leaves upwards of 800 million hungry. On 1 November 2019, Reuters published a story with the headline: 'Record 45 Million People in Southern Africa Facing Food Crisis in Next 6 Months'. This is the result of severe droughts, floods, and storms. In a region already accustomed to extremes, a series of unprecedented events is already putting at risk unprecedented numbers of people – and these numbers will only increase. There are other populations across the majority world facing similar survival-level challenges.

It is a stark and contemporary reminder of the unresolved trauma that underpins calls for global justice and reparations. The recognition and resolution of this trauma therefore needs to be at the centre of any new story. But like the existing story about dangerous anthropogenic climate change, the story about inter-continental redress has also failed spectacularly to emotionally connect and achieve resolution. Re-framing global justice so that it speaks to minority world vulnerability is therefore a key part of a new story.

A DILEMMA FOR XR?

Rebelling against the story as told is an existential question for XR. Not least because the pre-conditions for a new story aren't to be found in Roger Hallam's leveraging of the Holocaust as a comparative framing device, or in the Canning Town tube action during the October Rebellion. While the latter was admittedly focused on the more discreet demand of achieving economic disruption, it is the visceral display of dissonance that lingers. That dissonance being the image of two participants being dragged off the top of a train for expressing their vulnerability about the second half of the century, while the people on the platform below expressed their vulnerability about the next five minutes. None of this means the people on the platform aren't anxious about the future, even about the climate.

Taken together, the tube action and Roger Hallam's comments to German media neatly sketch out a dilemma for XR UK: meet people where they are and take them on the riskiest of journeys, or escalate tactics within an external story that's already contributed to a spectacular failure to act. With this dilemma in mind, increased public awareness of the climate emergency in the UK should be treated with caution.

We would go as far as to say that rising awareness of the Climate and Ecological Emergency might actually be shrinking XR UK's ability to cut through, post-April 2019. This was the take-away message of YouGov polling compiled for XR UK pre- and post-October's Rebellion, while anecdotal evidence shared on platforms (including Basecamp post-October) suggests action-attendance numbers are down and that co-ordinators are finding it hard to motivate their groups.

This appears to show our emergency messaging, as currently constituted, is the subject of diminishing returns when it comes to sustaining motivation and continually raising awareness. Recruiting rebels in 2020 is not the same as 2018/2019. It is set against the backdrop of mainstream media and the political class now bandying about the phrase 'climate emergency' with meaningless abandon. Power in effect has fed the in-built 'natural' psychological reluctance of the population to fully engage by conceding something to each of our three demands. In retrospect, the amazing achievement of our April 2019 Rebellion was to get the public and power to give lip service to those three demands. The hard work of getting real action – i.e., system-changing initiatives that come anywhere close to achieving the urgent and essential task of reducing emissions – is going to require far deeper engagement with a far more powerful story.

SO WHAT'S THE STORY?

In this pamphlet, we concern ourselves principally with the UK, but a Majority World-facing narrative is an equally important and urgent requirement. The new story will not be one about 'the environment' or 'green' issues. It will be human-centric and rooted in the indelible truth that we are living in a world that has evolved as if dangerous climate change did not exist. Bringing this realisation to life is our story as-yet-untold. It says we don't all go from now to extinction sometime after 2080 with nothing else in between. We lose everything that matters to us on the way: our public services, our security, our community, our homes, our food, our water. And ultimately the people we love. It is a story of unstoppable loss unless we act now. It is a story that starts with eliciting vulnerability.

Of course, vulnerability as a trigger emotion is not to be taken lightly. So as part of the research into this pamphlet, we reached out to Andre Spicer, Professor of Organisational Behaviour at Cass Business School. He in turn contacted a group of US academics whom he described as world leaders in research on communications by social movements and about climate change. Here is the key section in what he said:[93]

> This message of vulnerability has some important strengths: It triggers loss-aversion, a strong cognitive bias which tends to drive people to engage in more risky behaviour. It makes an abstract issue into a real issue through fore-fronting everyday issues like feeding a family. It brings the threats posed by climate change into the immediate time frame (5 to 10 years) which means they cannot be easily discounted away by people.

Professor Spicer also raises some caveats to the above:

> When people are made to feel vulnerable it can connect with powerful emotions associated with other times in their life they have felt vulnerable (such as childhood or traumatic situations). Although this can stir up strong emotions which prompt action, it can easily back-fire through prompt denial, reject or even anger. Experiences of vulnerability are used as a first step to get people to accept a new group or set of values. For instance, when recruits are socialised into a group they are made to feel vulnerable by having their prior identity stripped away. However, this is usually followed up by them being given a new identity through joining a group. This helps to make them feel less vulnerable.

As the professor underlines, it is dangerous for a new story to be consumed as a one-off vulnerability 'mind-bomb'. To leave people hanging

93 If you want to see his entire referenced feedback, goto Appendix 1 of https://xrstroud.org/wp-content/uploads/2020/01/XR-Story-Vision2020-Leaflet.V11.pdf

in a state of uncertainty and pain is to provoke denial and invite authoritarian forces to fill the resulting vacuum.

So how might XR manage these risks and tell a new story that fuses vulnerability to a vision for addressing inequality at the local, national, and global level? Well, in the first instance, the CEE needs to be repositioned as a lens exacerbating everyday vulnerabilities felt by people in relation to their gender, age, ethnicity, and class. Rooting the story in contemporary vulnerability helps connect the future with now.

Telling the story is all about embodied actions. That is to say actions that can carry a story. An element of the original genius of XR is that it is perfectly primed to do this, having fostered a network of hundreds of affinity groups, local and afar. With support and framing from the centre of XR, each group can deliver disruptive and non-disruptive actions that embody both vulnerability and a vision for a post-vulnerable world. Part of our vision for 2020, therefore, sees groups across the country simultaneously co-ordinating centrally framed actions to aid movement building and open up non-physical spaces for the new story to take root and embed.

Picture a contemporary sports stadium with the spectators holding co-ordinated coloured cards above their heads to create different patterns and messages. This is a metaphor for how actions could be both decentralised and choreographed to deliver shape-shifting mass impact and consciousness raising.

Such thinking is not without precedent: a year and a half before President Lyndon Johnson signed into law the landmark Civil Rights Act of 1964, more than seven hundred and fifty civil rights protests took place in one hundred and eighty-six American cities, leading to almost fifteen thousand arrests, demonstrating the enormous power of dispersed community-focused direct action.

As such, the remainder of this pamphlet will explore how actions can embody a new story about rising vulnerability and how it might be overcome.

SECTION 2: CAN YOU FEEL IT?

Hope is not the conviction that something will turn out well, but the certainty that something makes sense, regardless of how it turns out.

– *Vaclav Havel*

As the previous section set forth, XR needs to find a way of shifting our focus so that the system, which is making us all ever-more-vulnerable, is clearly perceived as the fundamental problem. XR's narrative and actions henceforth must bring out the failings of that system to keep us all safe.

So what follows is a proposal for evolving (not overthrowing) XR's present theory of change (ToC). We need, in the aftermath of last October's Rebellion, to make an honest assessment of the limits of the ToC we have been operating with so far. Within this, XR needs to consider that in 2020 it is still in movement-building mode on the journey towards system change. Put simply: we need to grow rapidly – not shrink – if we are to win!

XR needs to both change up the story and tell this story principally through its actions. The ambition is to help dramatically build the movement so that it can go on to mobilise the kind of numbers present in successful uprisings overseas, such as in the People Power revolution in the Philippines, as set out by the academic Erica Chenoweth.[94]

To do this, XR must first carefully distinguish between aspects of the academic research that apply to its purpose and those aspects that don't. Thus far, XR has sometimes displayed an unwise and undiscriminating reliance on academic research which does not necessarily transfer well from the domain where it was conducted to the domain in which it needs to operate. In particular, we need to take more seriously the point that Chenoweth's work (from which the ToC was

94 See especially her book *Why Civil Resistance Works: The Strategic Logic of Non-violent Conflict* (New York, NY: Columbia University Press, 2011).

grafted) does not necessarily apply to Western 'democracies', and that the oft-cited precedents of Gandhi, King, etc., may not be as relevant as assumed. This is because XR's aim is not to induct a discriminated-against group into full citizenship (as per 20th-Century struggles for civil rights and suffrage) but to realise system change so that all of humanity can live peacefully within planetary boundaries.

With the above in mind, there is a danger inherent in XR's strategy of aiming for 3.5% of the population, and equally in aiming at achieving X number of arrests and X number of people in prison. Movements like People Power in the Philippines *were hugely popular*, their getting up to 3.5% onto the streets was a symptom of this far greater popularity (a remarkable one, given the prospects of serious ill-treatment by the authorities). Yet that was not the main cause of their success, which was derived instead from huge public buy-in and increasing levels of defection from the regime. There is a danger that XR is creating the impression that if it gets 3.5% of the population 'on board', then it doesn't much matter if it alienates everyone else. This would be a false lesson to draw from Chenoweth's work. In reality, the successful movements she cites saw the 3.5% engaging in nonviolent direct action (NVDA) as only the visible part of an iceberg above the water. A majority of people – below the waterline – were not disconnected. Can the same, however, be honestly said of the British public when it comes to the CEE?

CORRELATION DOES NOT IMPLY CAUSATION

XR co-founder Roger Hallam noticed that successful rebellions tend to get a small percentage of the population taking part in illegal action, a far smaller number arrested, and a far smaller number imprisoned. He reasoned that if XR attained those numbers, then rebellion will be successful. But that simply does not follow. It is what is known as a fallacy of misplaced concreteness. Aiming directly at those numbers will fail if the actions undertaken in pursuit of the goal alienate much larger numbers.

Some of the above played out in October 2019. Numbers of arrests were up but this didn't create greater success. In fact, success was clearly lower than in April. Public attitudes towards climate didn't continue shifting in XR's favour in October. XR's own YouGov polling, mentioned in the previous section, clearly shows this, with XR's own popularity falling to 5% – a fall also indicated in the level of donations, email sign-ups, Facebook likes, etc.

This leads us to conclude that XR is already approaching terminal fatigue with its existing methods. Moreover, a vital part of a shift in strategy also needs to ensure that never again can a tiny unrepresentative sliver of the movement, in the name of the whole movement, undertake 'autonomous' actions hitting relatively poor communities[95] (however brave and well-intentioned that sliver may be).

The 3.5% figure did give some of us genuine renewed hope that real change may be achievable, but we have learned now that this tactic will not succeed on its own. Relying on this, if we don't diversify our offering, will lead to failure and burnout. Without abandoning the principles of mobilisation, it makes sense to foment, spread, and embody a story that will resonate with vast numbers of the public. That is the raison d'être of the story set out in the previous section. A chance to build a movement both wide and deep. We need to encourage the iceberg to form below the surface.

The public will only feel that it is genuinely sad/wrong/unjust that XR rebels are imprisoned if they deeply identify with the reasons that people sacrificed their freedom in the first place. This requires more than just a vague general sympathy; it requires being emotionally and/or intellectually on board. It requires story-led actions that make sense to the wider public so they understand and feel this as an emergency.

ONE MORE HEAVE?

We are sceptical that the current XR strategy can attain five thousand arrests on its proposed return to the capital's streets in 2020. Without a

95 See Chapter 18: those using Canning Town station early in the morning were primarily working class.

significant change of direction, getting even the same number of arrests as XR had in October 2019 will be problematic. We are also sceptical that five thousand arrests would overwhelm the justice system anyway. The US justice system has easily coped with similar – in fact with far greater – numbers in the past. Consider especially the May Day protests in 1971.

We think the 'one more heave' theory for 2020 is therefore flawed. It is premised on hoping something that is no longer working will work next time. For reasons set out in the first section, even if XR achieved its aim of five thousand arrests, the general public is (as things stand) unlikely to perceive this as anything other than a source of big irritation. Suffice to say, public sympathy for XR's cause will be negligible at best, and the sacrifice among those rebels arrested and imprisoned will be in vain.

You don't get sympathy by aiming for X number of arrests. You get it by doing something beautiful, powerful, intelligent, meaningful, and resonant, that challenges the authorities to either arrest you en masse – risking great public sympathy – or lets you get away with it.

So, using the shift in story suggested in the opening section, we propose we need to think hard about meaningful and resonant targets for actions. This is a step towards a story of change that could resonate more with rebels and with the wider public alike. Thereafter, XR will be in a position to reap the full benefit of the expertise it has developed around mass mobilisation.

In other words, what we are setting out here is a route by which we can achieve the goal of the original XR theory of change. A route by which we can arrive at numbers so huge, and popular sentiment sufficiently supportive, that mass mobilisation in the capital will be overwhelming.

VULNERABILITY AND ACTIONS

As outlined in the previous section, the story of our collective vulnerability is key. The way to tell it requires actions that embody and transmit that collective vulnerability.

Our new story could therefore be framed as reversing the (neo-liberal) gamble humanity made on efficiency over resilience. A first draft of this story appeared in the online magazine *The Conversation* in December 2019, by climate scientist Professor Will Steffen (lead author of the 'Hothouse Earth' report[96]) and systems expert Professor Aled Jones, who jointly highlighted how increasingly extreme weather events may soon become severe and frequent enough to cause what's called 'synchronous failure'.

This means a crisis in one country could – due to our interconnected global system – lead to failure in many others. Examples include (most crucially) food production, global supply chain resilience, political risk, insurance, and finance, to name but a few. In most instances, the just-in-time cost-minimisation philosophy applies, which means there is no resilience or buffer in the system. It's why supermarkets evolved to not hold any stock onsite. And why they'll be three-quarters empty within three days, if not re-supplied.

This calls upon the centre of XR to combine its understanding of climate science with that of complex human systems that sustain our everyday lives. In autumn last year, the BBC broadcast a three-part series called 'What Britain Buys and Sells in a Day'. It inadvertently showcased many of the vulnerabilities engineered into our food and manufacturing industries, which rely on a seamless, orderly, globally enabling environment to function.

The first step to designing actions to transmit these vulnerabilities would be for the centre of XR to reach out to systems experts, alongside the people who understand impacted industries and who can pinpoint the stories to tell. What, for example, will happen when that smooth enabling environment gives way? Of course, designing actions that embody vulnerability is more complex and inevitably carries risk. We are not advocating, for example, that groups of rebels across the UK simply head out and block a supermarket depot to disrupt supermarket deliveries across

96 https://www.pnas.org/content/115/33/8252

their region. Such an action would need to be very carefully designed and would need to embody recently leaked damning information (see below) highlighting the precise vulnerability, so as not to fall into the bear trap of previous 'head-turning' actions such as that at the Canning Town tube station. Rather, the type of embodied actions we envisage will flag up vulnerability, and will be nuanced, piloted, and built around inconvenience rather than outright disruption.

Decentralised and co-ordinated actions would therefore still affect working people but, if framed by XR correctly, wouldn't be overwhelmingly perceived as targeting working people in the same way that stopping public transport does. XR would instead be issuing a smart wake-up call. Front running Mother Nature, if you will, by disrupting in a relatively small but widespread way, now, so as to highlight and ultimately help prevent vast disruption in future.

ADDRESSING INTERESTS THAT MAINTAIN VULNERABILITY

Acknowledging that some of us will have to change way more than others – starting with the wealthiest – is part of embodying the new story. It is the polluter elite in the UK (and all other countries) who are most responsible for the climate and ecological crisis through their luxury consumption and their investments in polluting companies. The polluter elite – alongside other vested interests and central government – can and should be positioned by XR as blockers to the resolution of vulnerabilities that will be exacerbated by climate change-induced shocks.

The polluter elite are billionaires and their multi-millionaire kin, who profit the most from the economic system that's destroying the planet. We envision XR making this visible in a different way to what has gone before. This is vital, since previous attempts by environmental NGOs and other civil society groupings have largely failed.

One possible way is for XR to leverage the existence of TruthTeller.Life, an online gateway for whistle-blowers – developed by XR in 2019 – to tell their truths about the future through the anonymous disclosure of confidential information or leaks. If supported by embodied actions, TruthTeller.Life could help realise XR's first demand by encouraging truth-tellers working inside the system to become invisible rebels by leaking withheld information.

Above all, it's essential that actions embody the new story by also calling in rather than calling out the targets of planned actions. In essence, be loving as well as raging, which is not a big stretch for XR, as we have consistently, successfully achieved this in the manner of our protests (with only a few exceptions that have drawn disproportionate media attention). This pivot to a new story will be essential for movement-building in 2020, as XR seeks to shrug off the tag of being 'elitist, middle-class, and out of touch'. The intention is to demonstrate we are on the side of working people without coming across as ideologically motivated class warriors. This underscores the shift in storyline and storytelling through embodied actions.

It says: if you're a believer in social justice, and deeply concerned by the real threat of climate breakdown, be part of XR. If you would like capitalism to evolve beyond its destructive tendencies and are deeply concerned by the real threat of climate breakdown, be part of XR. But you don't have to be either, or anything else. You just have to believe in nonviolence and the need to act now because of the climate emergency. The ambition is for us as a society to really feel this emergency at last. If the polluter elite go on as they are, then it's curtains for humanity. That is not ideology. It is plain and simple fact. This is what the somewhat misleading slogan 'beyond politics' really means: that very radical action is now needed in order to enable us to hold on to what we've achieved. That action will involve the creation of a more equal society – not for reasons of ideology, but for reasons of survival.

A VISION OF POST-VULNERABILITY

It's incredibly difficult to face up to this awful climate reality even as it becomes more obvious all over the world. It's also going to be incredibly difficult to change this system, for the reasons set out in this section and the previous one. It is why the new story also holds a vision for what the future might look like when today's sources of vulnerability are resolved.

For the reasons just set out, we believe this is rooted in visioning a significant shift towards equality at the local, national, and global levels. That shift is part-symbolised by the (non-rancorous) targeting of the polluter elite at each level.

But the highlighting of contemporary vulnerability is in effect just the nose of the new story. It is why, in 2020, we envisage XR elevating the relevance of adaptive and regenerative measures at local, national, and global levels. This invites us to consider how XR might embody a vision for a post-vulnerable world and is the subject of our closing section.

SECTION 3: CAN YOU DREAM IT?

The future belongs to those who believe in the beauty of their dreams.

– Eleanor Roosevelt

As the opening section set out, XR needs a story – and a broader invitation to act – that speaks to non-environmentalists. We therefore envisage a Rebellion that appeals to social, emotional, and cultural yearnings. One that not only links people to their core values spanning respect, fairness, morality, and love, but one that connects the personal with the political – or societal – backdrop that holds so many people in check.

The phrase 'shit life syndrome' is now a commonly used term for that backdrop among the nation's general practitioners. It describes conditions of patients arising from poverty and record levels of loneliness and isolation, now recognised as a leading cause of disease and death. Indeed, our separation from ourselves and each other seems only to accelerate in line with the speed at which humanity surpasses planetary boundaries.

But it's really beneath such statistics that XR needs to focus: on the everyday and unsung stories of disconnection and vulnerability that might not be the stuff of headlines, but are nevertheless deeply held and painfully felt. At first glance, this can feel very different from the defiant and visible energy of rebels sounding the Climate and Ecological Emergency.

A TALE OF TWO INVITES?

Both these energies were arguably present during October's Rebellion when, alongside the dominant energy of occupation and disruption, a parallel story emerged about a deep awakening to the reality and pain of the violence and harm humans are causing to each other and the natural world. This story took the form of a 25,000-strong 'grief march' that snaked its way through the rain-soaked streets of London's West End on the middle Saturday of Rebellion.

For some rebels camped out in the cold and damp for five days already, it probably looked like just another march that wasn't going to change anything. But then there can be no shrinking from the fact that 3.5% of the population (from which we shall assume the marchers were drawn) need another way to collect and express their solidarity. This was a regenerative, movement-building action that contributed to our second principle and value (P&V):[97]

Principle 2: We set our mission on what is necessary. Mobilising 3.5% of the population to achieve system change – using ideas such as 'momentum-driven organising' to achieve this.

But perhaps this unearths a creative tension – or even a paradox – with our third P&V:

Principle 3: We need a regenerative culture. Creating a culture that is healthy, resilient, and adaptable.

97 For a listing of our Principles and Values, see https://sites.google.com/site/empathycirclehub/projects/xr-principals

When we say we need to mobilise only two million people, how does that connect with building a regenerative culture internally and externally through our messaging, storytelling, and actions?

Because a healthy, resilient, adaptable culture depends on *most* of us being in healthy relationships with one another and the earth. We can't live in a microcosm. Any regenerative culture we pursue or cultivate has to live in deep empathy with those that it comes into contact with.[98] A culture that cultivates self-awareness, inner knowledge, and agency, as it seeks to engage and collaborate across differences and divides. A culture of belonging and community connectedness where each community supports values of respect and kindness towards the 'other', undivided from the natural world. This is about leading an undivided life that recognises the pluriversality and complexity of all life and its ecology. It needs everyone to feel acknowledged and appreciated. Our third P&V demands we communicate with more than 3.5% of the population.

In contrast, XR's second P&V evokes what the late American cultural anthropologist Margaret Mead had to say about social change: 'Never doubt that a small group of thoughtful, committed citizens can change the world; indeed, it's the only thing that ever has.'

Has such thinking inadvertently given some rebels permission to close their ears and hearts to the majority of the general public, because supposedly the latter aren't needed? Has this in turn created licence within XR for alienating working people and people of colour? Put another way, how did the Canning Town tube station action fit with XR wanting to be part of birthing a paradigm shift where people are able to embrace interdependence and live in regenerative cultures? Are we in fact telling ourselves that XR can co-lead a march to radical equality, reconciliation, and deep collaboration, at the same time as it separates itself from the perceptions, stories, culture, and needs of 96.5% of the

98 Skeena and I discuss this aspect in much more detail in this 'empathy circle', which is itself a fascinating process worthy of wider understanding: https://www.youtube.com/watch?v=2zfzhjR2LLg

population? If so, are we back to trying to operate on thin ice without a solid iceberg beneath us?

Our full ToC recognises that we are here to create a polarisation phenomenon. So yes, in its own way XR is necessarily a divider and, given its tactics, has and will continue to be perceived by some as confrontational and unreasonable. Perhaps a through line for XR is offered up by what authors Mark and Paul Engler had to say in their book *This is an Uprising*: 'For polarisation to pay off, the positive must outweigh the negative. And here the reaction of the general public – those not already aligned with either side – is critical. When the process works, members of the public are alienated by the extremism of reactionary opponents.' This means that we are counting on a broad base of sympathy at the same time as unreasonable reactions from the state. Building this base is going to be enormously hard for XR in 2020 when the means of communication are so tightly controlled and manipulated by a few organisations with gigantic systemic power.

Thus, XR also needs to look at an engagement strategy that works with frustrated journalists, advertising agencies, and businesses, aware or open to the challenge of circumventing mainstream media while also trying to fence with it. This is where centrally framed co-ordinated actions can also add value, because it will be much easier for people in the UK to find sympathy with the purpose of XR if both the story and the invitation to participate in nonviolent action are local.

WHO ARE THE 3.5%?

So far, XR hasn't chosen to specifically target a certain demographic in terms of rebel recruitment – the magic 3.5%, so to speak. What XR has observed – and been lambasted internally and externally for – is that most rebels are white and middle class, using their power and privilege to demand change. XR can be enormously grateful for the efforts of such rebels while taking credit for providing them a unique space to come together and collectively embody their anxiety about the future through loving and courageous discussions and actions. The quadrillion-dollar question is how to scale up this precious invitation and make it more diverse.

Right now, as demonstrated by XR's current strategy process (ongoing at the time of writing), growing that invitation is frustrated by a consistent lack of space, recognition, and championing of diverse ideas and voices in the centre of XR. In its place, power is wielded through structures, co-founder power, strategy and tactics formation by a very small group of people.

Zack Exley and Becky Bond reflected on the failures of the last Bernie Sanders campaign in their book *Rules for Revolutionaries*. They dedicated a chapter to their conviction that revolutionary campaigns must be co-led by the most marginalised voices in society, not least because such people are able to better see the blindspots that oppression and prejudice garner for those benefiting from systems set up to serve elite privilege. We feel XR would do well to take this on board, so as to genuinely embody our seventh P&V:

Principle 7: We actively mitigate for power – breaking down hierarchies of power for more equitable participation.

And maybe this P&V is a gateway, as it's only by committing to the above that XR can genuinely embody the following P&Vs:

Principle 1: We have a shared vision of change

Principle 4: We openly challenge ourselves and our toxic system

Principle 6: We welcome everyone and every part of everyone

A shared vision of change literally needs to be shareable and relatable, which requires an inherent and identifiable commitment to representation and redistribution of power. To challenge our toxic system, we need to challenge the oppression and dehumanisation of those that don't fit into white privilege and be honestly welcoming to everyone. We need to deliberately and consciously make space for those that are different from us.

This could also naturally relieve any tension between our second and third P&V outlined above – that we need an early small minority of people (3.5%) to support our ask alongside building mass awareness and empathic connection among the wider population. It tells us that the complex and comprehensive work of reconnection and reconciliation cannot be shunned. It is the process by which our inner work lights our shadows to connect with the outer work of taking us out of our separation and powerlessness and into regenerative cultures. It's where we know we have to decolonise our attitudes and behaviours if we are to enable planetary repair.

The requirement for reconnection and reconciliation (R&R) cannot be underestimated. Below are six levels of R&R work that XR could choose to embody through storytelling, networking, and actions. It is our R&R work that will enable our bold and beautiful visions to begin to create power and energy enough for transformation.

1. Inner reconciliation – reconciling with our truth, grief, fear, and rage with self-care.
2. Community reconciliation – supporting community cohesion and belonging.
3. National and global reconciliation – supporting the reconciliation of divided communities and facilitating reconciliation work between and within over-consuming and lower-consuming countries.
4. Earth reconciliation – encouraging renewed relationship with and respect for non-human life and earth elements (air and water).
5. Reconciling with those we view as different or as 'other'.
6. Reconciling and encouraging relationship with greater consciousness or that which we don't know through our minds.

Through our principles and values and vision, we wonder if we can model what the future is asking of us all by being genuinely prefigurative.

If we can lean into the creative tensions and conflict outlined above, and apply new ways of transforming what is violent, difficult, and stuck, perhaps we can we make a genuine contribution to the work of R&R as a precursor for building truly regenerative cultures. If XR seeks to be the transformation it is asking of the world, then we might catalyse the pre-conditions for a leap in empathy consciousness required to deliver on XR's second demand, which is tantamount to system change.

With this in mind, below is a diagram, included in our original DNA training for rebels. This Venn diagram is based on the thinking of Gandhi, as interpreted by Chris Moore Backman and Joanna Macy, founder of 'The Work that Reconnects'. It has been adapted for XR to guess at what the most enormous change in human history might involve.

The diagram suggests that we cannot hope to secure the vast transformation of systems and relationships that we need without three strands of inner and outer action happening all at once. We need personal, local, national, and global change. Harm and violence exist in all three domains and we simply cannot change one without the other as they all trigger and feedback on one another.

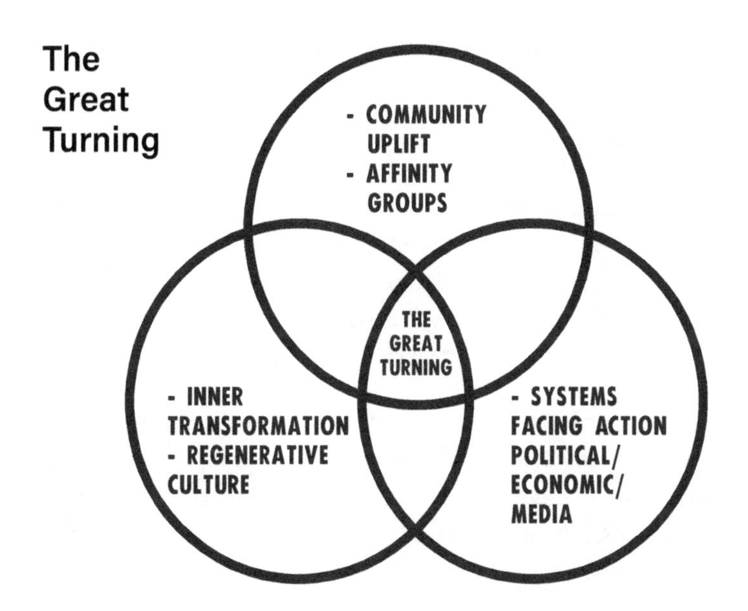

In a UK context, the very opposite played out during the December 2019 General Election. It illuminated a mass sense of powerlessness where the highly manipulative messages 'get Brexit done' and 'take back control' successfully attracted 45% of voters. Powerlessness, as discussed by international teacher Miki Kashtan from the Nonviolent Global Liberation Community, is one of the components of Toxic Patriarchy. The other two being the Separation and Scarcity stories. Combined, they have arguably been the origin of colonialism, class war, racism, militarism, and, most recently, neoliberalism.

We all carry this story in our DNA. Ultimately it is the reframing of this story, and the rewiring of the heartache it feeds, that will generate the reconciliation and collaboration now necessary for deep adaptation and resilience.

So how can XR go about dreaming, suggesting, and embodying such a vision for a post-vulnerable world? To date, our commitment to emergency-mode messaging has been almost exclusively focused on one of its three stages set out by clinical psychologist Jane Morton:

1. Tell people there is an emergency.
2. Tell people change is possible: inspire them with a vision of change.
3. Ask people to act according to their values.

Evolving our response to stage one (as per sections 1 and 2) and extending into stages two and three means XR committing to a very different kind of polarisation. It means having the courage to divide people from business-as-usual with a courageous story of our collective vulnerability and our collective moral instinct to love and respect.

In accordance with the views of Professor Andre Spicer mentioned in the opening section, it means being sensitive about evoking a fear that sends people into a despair that further fuels a paralysing sense of powerlessness. Rather, we need to deploy a generous and emotional story about what makes us all vulnerable and equal in our drive to move from surviving to thriving. We can remind people that all life wants to grow and thrive and that life can adapt to the harshest conditions. That we are all necessary. That we can dare to be great, as the late Polly Higgins, lawyer and ecocide campaigner, used to say. Rooted in truth and love, drawing on our nobility and grace, acting with fierce courage and not turning our face away from the suffering already in high command, we can move through this moment. We can do what is beyond the furthest lights of our imagination but within all possibility.

This means telling – and remembering and feeling and modelling – stories of togetherness, trust and empathy. It means describing the known and imagined beauty of radical equality and resilience through word, image, sound, and performance.

So instead of bowing to the 'scarcity' story, we share a vision of 'abundance' that naturally resolves our vulnerability. Not an abundance of stuff and things, but an abundance of what is vital to life: good food, good community, living in congruence with our values, in honest relationship with the natural world, everything that gives meaning.

Inherent here is a shift in energy from only regarding ourselves – or being typecast – as activists driven by 'impossible demands'. It suggests a more gentle or nuanced energy for also becoming activators that will be essential for movement building and systems transformation. It is, if you will, #BeyondActivism.

This may sound clichéd to those of you who've been involved in politics, but this shift will begin in earnest from a place of listening to and understanding the public. It will be working with a movement of movements, including UK equality networks, to listen for the relationship and empathy needed from XR. To build a mass movement, XR will need to create a relatable story for different segments of the public by borrowing from the best that political communications has to offer. A story that means something to people who will not suddenly care about man-made climate change because there's another forest fire overseas, or a domestic flood 200 miles away.

The same people, however, care about and are interested in doing 'the right thing' by the people they love and want to protect. Just like the alienated commuters on the platform at Canning Town tube station. XR needs to tell a story that activates these instincts that lead to the withdrawal of consent from the bankrupt and broken systems that are destroying what people love, or at least sympathy with those people in active rebellion.

REGENERATIVE ACTIONS

The central theme of this pamphlet is that XR needs to not only tell the new story but embody it through actions. In 2020, XR needs to include regenerative and restorative action so that our lived reality can pivot from vulnerability to radical equality. It de facto means thinking more carefully and creatively about how actions might carry both the story of vulnerability and our vision for its resolution. Given such actions will be XR's primary means for movement building, creativity is a must alongside a broader invitation for participation.

Actions that unfurl a vision for the future might tell a story of regeneration, rewilding, and repair. One idea, contributed by Professor Jem Bendell (author of 'Deep Adaptation: A Map for Navigating Climate Tragedy', the most downloaded climate change paper of all time), sees XR affinity groups occupying popular local green spaces to plant fruit and nut trees and vegetables. A pop-up allotment so to speak. The action could be co-ordinated nationally and be timed to be part of other actions around vulnerability, as suggested in the previous section.

Regenerative actions might also start with a mass sing or choir and end the same way. How would it be if XR called Friday or Sunday Assembly actions (as well as citizen assembly actions) around reconciliation? Or, as already has been suggested elsewhere, XR could invite rebels to come together and volunteer for clear-up operations in natural disaster areas or food bank support work.

The authors of this pamphlet aren't expert at designing mass-coordinated actions or local engagement. The ideas we've outlined here are merely intended to help cajole and, if we're fortunate enough, inspire those already working at the centre of XR UK, alongside many thousands of rebels across the UK, to take on the story and make it their own. As such, we envision the movement's regenerative cultures and visioning circles working in a more integrative and collaborative way, with XR Communities, XR actions, XRLiberation, XRISN and XRIST.

GOING GLOBAL

While this pamphlet is not without ambition when it comes to challenging XR, perhaps the biggest ask of all is that a new story needs to extend far beyond the UK. XR's growing international network needs to be part of developing, evolving, and tailoring the story across numerous countries. It is a recognition that the future of everyone in the UK will depend on global responsibility, will, and solutions. Thus XR needs to inspire global visioning that tells a story that further ignites global NVDA, as part of the telling of our interdependence as a source of near-term vulnerability but ultimately longer-term reconciliation. That is to say, a very different vision for globalisation.

In conclusion, XR needs to live and breathe its principles, values, and vision in a more synthesised and coherent form. The value of the power we have created lies in us helping to birth a tremendous collective change of heart that sits at the core of our P&Vs and Vision. It is why XR now needs to build greater solidarity and coherence especially with its international groups, so that it can go on to embody a global story that disrupts international institutions and their toxic pillars. Whilst we cause disruption, we need to kindle togetherness so that we might also inspire the greatest conflict transformation and peace process the world has ever known. We are holding a most beautiful paradox. A place where love protects and says no and a place where love says yes and shows its infinite care.

BACK TO YOU...

Congratulations if you made it this far! As the title of this pamphlet laid bare, we think XR risks rushing its response to the 'emergency' in

2020. It sounds the most incongruent of phrases, we know. But we see no substitute or shortcut for the more patient demands of storytelling and movement building by way of embodied actions. In particular, we point to the courage and commitment needed from all of us set out in the closing section.

We need to diversify the energy that propelled XR to success in April 2019 by way of creativity, shapeshifting, and visioning. How else will we embody a resonant story that not only teases out vulnerability but also embodies humanity's journey towards equality at all levels?

In the vast transition that is coming, everything will have to change. But those that will have to change the most are those with the most. Those who have most responsibility for making us all vulnerable need to feel that sense of crisis that we help create. Eventually, we can come back to London and take to the streets in such huge numbers, and with such popular backing and understanding, that the authorities are simply overwhelmed.

That – getting our demands met – is the prize that getting our story right could yield.

RESOURCES

EXTINCTION REBELLION

Bradbrook, G., 2018, 'Heading for extinction and what to do about it'. (18 September 2018). Available at: https://www.youtube.com/watch?v=b2VkC4SnwY0

Extinction Rebellion website: www.rebellion.earth

Extinction Rebellion, 2019, *This is Not a Drill: An Extinction Rebellion Handbook.* Available at: https://www.penguin.co.uk/books/314/314671/this-is-not-a-drill/9780141991443.html

Extinction Rebellion, 2019, 'We have three demands. #1 Tell the Truth, #2 Act Now, #3 Beyond Politics' (21 April 2019). Available at: https://www.youtube.com/watch?v=4s9jfkWz72g

Extinction Rebellion, 2019, 'We Have the Right to Rebel: Extinction Rebellion Begin Westminster Shutdown'. *The Guardian.* (7 October 2019). Available at: https://www.theguardian.com/environment/video/2019/oct/07/extinction-rebellion-mass-protests-restart-london-video

Extinction Rebellion, 2019, 'BBC Politics Live | Dr Rupert Read | Extinction Rebellion'. BBC Politics Live. (9 October 2019). Available at: https://www.youtube.com/watch?v=Zl0qMuF6ec0

Hallam, R., 'Common Sense for the 21st Century'. Available at: https://www.rogerhallam.com/wp-content/uploads/2019/08/Common-Sense-for-the-21st-Century_by-Roger-Hallam-Download-version.pdf

Morning Star: article looking back over the first year of XR https://morningstaronline.co.uk/article/f/we-could-get-20000-or-30000-people-willing-take-direct-action-streets

Timeline of XR actions: https://en.wikipedia.org/wiki/Timeline_of_Extinction_Rebellion_actions

Truth-teller – anonymous disclosure of insider information. Access website at: https://truthteller.life/

CLIMATE BREAKDOWN AND WILDLIFE LOSS

Alliance of World Scientists, 2019, 'World Scientists' Warning of a Climate Emergency (14 October 2019). Available at: https://scientistswarning.forestry.oregonstate.edu/

IPBES, 2019, 'Global Assessment Report on Biodiversity and Ecosystem Services' (6 May 2019). Available at: https://ipbes.net/global-assessment

IPCC Special Report, 2018, 'Global Warming of 1.5°C' (8 October 2018). Available at: https://www.ipcc.ch/sr15/

NASA, Global Climate Change: Vital Signs of the Planet. Available at: https://climate.nasa.gov/vital-signs/ice-sheets/

National Biodiversity Network, 2019, 'State of Nature Report' (3 October 2019). Available at: https://nbn.org.uk/wp-content/uploads/2019/09/State-of-Nature-2019-UK-full-report.pdf

WWF, 2018, 'Living Planet Report' (29 October 2018). Available at: https://wwf.panda.org/knowledge_hub/all_publications/living_planet_report_2018/

RUPERT READ

Rupert Read personal website: https://rupertread.net/

Read, Rupert, 'This Civilisation is Finished: So What Is To Be Done?'. Talk at the University of Cambridge. Video of talk available at: https://www.youtube.com/watch?v=uzCxFPzdO0Y

Sunday Times profile of Rupert Read available at https://www.thetimes.co.uk/article/frightfully-sorry-but-ill-have-to-make-your-drive-home-a-horror-qjvs36v56

Taleb N., Read, R., Douady, R., Norman, J., Bar-Yam, Y., 2014, 'The Precautionary Principle with Respect to Genetic Modification of Organisms'. Available at: https://arxiv.org/pdf/1410.5787.pdf

SAMUEL ALEXANDER

Samuel Alexander personal website: https://samuelalexander.info

Simplicity Institute website: https://simplicityinstitute.org

OTHERS

Bendell, J., 'Deep Adaptation'. Available at: https://deepadaptation.info/

Wallace-Wells, D., 'Climate Change and the Future of Humanity'. Five-minute animation available at: https://www.youtube.com/watch?v=eUh-TXKldiE

The Precautionary Principle – Climate Change debate animation: https://www.youtube.com/watch?v=MfsVlkaExF8

CPSIA information can be obtained
at www.ICGtesting.com
Printed in the USA
BVHW030101060822
643962BV00010B/726